QUANTUM MECHANICS THAT MAKES SENSE

Louis S. Jagerman, MD

Table of Contents

An Introduction

Contemporary science, and theoretical physics in particular, are reputed to be incomprehensible. The main "offenders" are relativity and quantum mechanics. Relativity is thought to be reserved exclusively for Einsteinian geniuses, while quantum mechanics is said to be unbelievable, illogical, and reminiscent of black magic dubbed "quantum weirdness."

Relativity "is not so bad," thanks to a large body of explanatory literature that sprouted over a century ago. Ever since, many authors have managed to clarify the principles of relativity and to dispel the notion that the topic is beyond the grasp of ordinary mortals. For instance Einstein himself, the "father" of relativity, wrote clarifications for non-scientists, as did many others, and I have contributed to this effort.[1] Moreover, elements of relativity—such as "space-time" and the key equation $E=mc^2$—have become part of everyday vocabulary and often are included in basic education. Indeed, once a handful of underlying principles is grasped, both special and general relativity turn out to be quite rational and logical, instigating few controversies and engendering no serious conflicts with earlier "classical" (Newtonian) forms of physics.

Briefly defined, quantum mechanics is modern subatomic physics, but it presents greater obstacles than does relativity, as it implies that physical reality is not rational. This branch of physics is also younger, as is its literature. Moreover, quantum mechanics is a compilation of the ideas of many "fathers," some of whom disagreed vehemently with each other, and even today sophisticated scientists argue over various "interpretations" of quantum mechanics. Meanwhile, popular writing on quantum mechanics stresses principles that seem to be incompatible with our traditional and apparently rational scientific truths. Paraphrasing such approaches, "you should not expect quantum mechanics to make sense," or "if you think you understand quantum mechanics, something is wrong with you."

Of course controversy has surrounded many advances in science. Thus the relativistic notions of "curved" time, curved space, and—most notably—curved space-time initially were met with skepticism and incredulity, mainly because they defied common sense and were not evident in ordinary experience. However, today even the deepest principles of relativity are so familiar that almost all scientists accept them without hesitation. Indeed, as I will discuss, some facets of relativity are essential to contemporary quantum mechanics. The difference is that even sophisticated minds still find many of the basic tenets of quantum mechanics to be irrational and to include preposterous paradoxes and enigmas that resist full explanations. In short, quantum mechanics appears to contain "mysteries."

My thesis is optimistic. We need not surrender to the idea that quantum mechanics will never make sense. I maintain that if we accept only a few assumptions—my postulates—these mysteries can be resolved, and the entire topic will cease to be intellectually so forbidding. Despite its apparent mysteries, we can continue to appreciate the great virtue of quantum mechanics: It indeed can elucidate many basic phenomena that cannot be explained or predicted by classical physics or by relativity, and its application leads to correct mathematical and experimental predictions. In short, even if it seems enigmatic, quantum mechanics works. My approach is to identify the most fundamental and simplest scientific concepts that can explain the various mysteries of quantum mechanics, the ones said to be unfathomable. In particular, I will present modern ideas in physics which can serve as my postulates in order to provide a logical framework for quantum mechanics. I will show that such ideas do exist, that they are reasonable and rational, and that they have manifestations which, to a lesser or greater extent, can be actually observed. I shall explain these ideas and my postulates in detail and apply them to my theme, which is to dispel the mysteries of quantum mechanics.

Of course an important step is to list these mysteries. The numerical order in this list is roughly governed by traditional thinking, and such lists appear routinely in scientific literature. Here is a typical (but not the only possible) hierarchy, which obviously implies that wave-particle duality is the most enigmatic item on the list.

---

[1] See Jagerman in Bibliography starting on page 71.

1. Wave-particle Duality
2. Entanglement
3 The Uncertainty Principle
4. Quantization
5. Tunneling
6. Technical Micro-mysteries
7. Fields in Quantum Theory

One of my concerns must be the reader's knowledge of quantum mechanics. To ensure sufficient familiarity I will dwell briefly on the basics for each member of the above list. To help elucidate each item, I will also adjust the order in this list to match my priorities. Quantum mechanics raises many questions, but I believe that once the main mysteries have been demystified—as I plan to do via my postulates—the bulk of the topic will be clear, and lesser issues, not listed here, will require no elaboration. Readers advanced in quantum mechanics are asked to keep these goals in mind, for I will skip most of the minutiae and practically all of the mathematics, and I will abbreviate many topics so as to maintain the focus of this publication. I am aware—and I emphasize—that I am not addressing the professional scientist.

As a preview, the key mysteries whose resolution is essential are these, in the order that befits my approach:

1. The Uncertainty Principle
2. Wave-particle Duality
3. Entanglement
4. Quantization
5. Tunneling
6. Technical Micro-mysteries
7. Fields in Quantum Theory

Hence my dissertation has seven main sections. In each section I will discuss one of the above "mysterious" components of quantum mechanics and then, where appropriate, I will present postulates by which I will dispel the mysterious quality of that component. However in the case of the first listed component, the uncertainty principle, I prefer to start the section with my postulate and then take up that component; in this way the logic will flow better. When I get to the last items in this list, separate postulates will not be essential, since a few additional broad principles answer most questions. My themes will frequently intersect, and the first section is...

The Uncertainty Principle and Jiggling (with a note on Chaos)

As I said earlier, I prefer to cover the appropriate postulate first, the one I have named THE JIGGLING POSTULATE. The underlying hypothesis for this postulate is this:

The smallest constituents of the universe are in constant delicate vibrational random motion.

I call such motion "jiggling," and though it does not sound like a serious scientific term, I will use "jiggling" as a keyword. By "smallest constituents" I mean subatomic and somewhat larger particles. In theory, everything in the universe must participate in this jiggling, but jiggling is too subtle to be detected or to be significant in objects larger than about the size of a molecule. Conversely, the smaller the object, the more conspicuous is its jiggling.

Jiggling is "constant" in the sense that particles are always jiggling. This means that the essential characteristics of jiggling do not change over time, nor do they vary from place to place. In particular, I assume that jiggling is basically the same everywhere in the universe, that it has been this way for a long time—perhaps even since just after the "big bang"—and that it will not change in the foreseeable future.

On the other hand jiggling is "delicate" in the sense that it is easily disturbed. "Fragile" is an alternative adjective for this property. As an analogy, the flickering of a small flame is so delicate that even shining a weak beam of light on the flame visibly disturbs its flickering. In other words jiggling can be altered, and I will show that this possibility is important.

I further describe the fragility and "delicateness" of jiggling as follows: I envision the jiggling of any one particle to be a very low-energy process, though the exact amount of energy at any moment is unpredictable. (Energy participates in jiggling, and I will return to this concept.) Part of the "delicateness" or fragility of jiggling resides in this low-energy nature, as revealed in its elusiveness. I also envision jiggling to be a kind of very low-amplitude vibration, but the exact amplitude is also unpredictable. The low-amplitude nature of jiggling also adds to its fragility.

Finally I envision jiggling to have very high but unpredictable frequency. Yes, frequency and energy are linked; if energy were completely constant, jiggling might be only a one-frequency hum (quantum harmonic oscillation). However, in line with the energy-time uncertainty principle, which I shall detail later, the energy in jiggling is liable to very short-lasting but random fluctuations, reflected in its unpredictable frequency.

These properties—low but unpredictable energy, low but unpredictable amplitude, and high but unpredictable frequency—account for why jiggling is difficult to observe directly. Jiggling is further obscured by (1) size, such as in anything larger than a molecule, (2) other forms of diminutive motion, such as thermal vibration, and (3) the effects of the acts of observation and measuring, such as are done in experiments that I will discuss later. Please note the common feature of these properties: Unpredictability!

Jiggling is "vibrational" in the sense that it is a very rapid three-dimensional quivering or shivering or trembling or shaking. This activity is separate from other gross forms of motion, notably travel in some overall direction, such as falling, orbiting, or being thrown or emitted. These kinds of motion, of obvious interest in general physics, are what I call directional motions, while others call them "longitudinal," and they often are the consequence of a measurable external force. Directional motion need not be straight-line—for example falling in a trajectory and orbiting are cases of curved directional motion—but it must involve moving something, such as a particle or object, between two significantly distant locations. Thus, if a photon moves from a lamp to an object across the room, I consider this to be directional motion, though something as small as a photon also has significant jiggling. In every-day cases jiggling is too subtle to affect directional motion, but jiggling does accompany it, and in theory (per Cox and Forshaw, p. 46 [see Bibliography, page 71]) the photon later "can be anywhere else in the Universe...."

The crucial feature of jiggling is that it is "random," so that no matter how we dissect and analyze jiggling, we will always encounter an element of unpredictability. In other words jiggling entails some unavoidable *probability.* What is particularly unpredictable (probable) about jiggling is its direction. Moreover, as mentioned above, frequency is tied to energy, and later I will explain how energy may fluctuate randomly, so that the frequency and amplitude of jiggling are also unpredictable. (My use of jiggling as a decisive physical process can be called a "stochastic" perspective. Stochastic mathematics studies intrinsically random systems.)

Indeed randomness justifies the entire notion of jiggling, and here is why: I posit that any measurement of a physical process or event that is affected by jiggling is also subject to a minimum but inescapable degree of unpredictability. As I already emphasized, in practice only submicroscopic events reveal jiggling, so that most ordinary physical measurements are not appreciably affected. In any case, a key point is that *in theory no physical law, principle, rule, equation, or generalization exists that can make the effects of jiggling fully predictable.*

Meanwhile, all physical process and events entail various particles, including the fundamental subatomic particles. This means that in theory we cannot calculate or otherwise foretell the exact effects of jiggling on the interaction between particles. Again, this does not mean that jiggling is always clearly evident; in fact it is rarely discernible, but in principle it is unpredictable as well as ever-present. On the other hand the features I ascribe to jiggling, particularly its fragility, allow two or more particles to jiggle in some degree of harmony. The possibility of such harmony will enter into my discussion of "entanglement."

This brings me to a functional definition of jiggling: Imagine that an object has been reduced in size to one particle, and that this particle is of the smallest possible variety. Further, please imagine that it has been frozen down to absolute zero, that it is completely isolated from all energies and forces, and that we have done everything possible to keep it from moving. *The sole remaining activity of such an object is jiggling.* Of course such a circumstance is highly exceptional and hypothetical, but it is also very simple; we need to know very little about the other forms of motion or about forces or external energies, except to assume that they are absent here and now. The motion we cannot halt and that is never fully absent is jiggling—random, constant and fragile but ineradicable jiggling. (Jiggling in not directly associated with the background radiation in the universe; the latter is thought to be a remnant of the big bang, but it is not essential to my thesis, nor is background radiation of any kind necessarily random.)

Obviously, in my view jiggling is a basic and essential element of physical nature. On a practical level, as I will discuss in detail later, once we apply the idea that this kind of intrinsic unpredictable vibrational motion is responsible for the uncertainty principle, this principle and other main tenets of quantum mechanics will cease to appear mysterious. In short, my position is that *mysteries of quantum mechanics are explainable if we think of this jiggling as a vital constituent of physical behavior.*

A legitimate question as this juncture is whether jiggling really exists, and in particular whether it is any different from the familiar and more obvious thermal vibration of atoms and molecules. My answer is that jiggling itself is a separate entity, appearing as an independent property of all particles. Therefore the interplay between jiggling and temperature deserves attention. Thermal motion is a vibration of the particles in an object that has temperature. This vibration is largely random, but this randomness is of no direct pertinence in my thesis. Thermal motion has no overall direction, except in the sense that one location may be hotter or cooler than another nearby location, so that a movement of heat occurs until a thermal equilibrium is reached. Sufficiently vigorous thermal vibration reveals itself as measurable and potentially destructive heat, whereas jiggling poses no threat. (Brownian motion is also very temperature-dependent.)

More importantly in this context, since everything in the universe has some temperature—since nothing in nature is at or below absolute zero temperature—every particle has some thermal motion. Of course particles in hot objects have more vigorous thermal motion than in cold objects, and if an object or particle can be cooled to absolute zero (which is zero degrees on the Kelvin scale, –273 on the Celsius scale, and about –459 on the Fahrenheit scale) all thermal motion must cease. However, getting even close to this limit is extremely difficult.

Thermal and directional motions are also independent of each other, but (said again) even if a particle has as little thermal motion as possible and no directional motion, jiggling motion does not cease. I stress that zero temperature does not equate to zero jiggling. Very importantly, this also means that *some element of probability must persist in the behavior of all matter (and energy) in the universe at any temperature.*

Another legitimate question is what sustains jiggling. I approach this issue by focusing on energy and on energy "fields" rather than on motion of particles. (Fields in quantum theory have measurable properties at each point; more on this in a separate chapter on fields, starting on page 62.) Zero absolute temperature means zero thermal energy, but we know that there exists a "zero-point energy" ("ZPE") or a "ground state energy." This energy is discernible as the random electromagnetic oscillation that is left in a vacuum after all other energy has been removed. Also, an irreducible amount of energy is present in a field from which all matter has been removed and nothing remains that can move.

An energy field in these conditions is a "zero-point field" in the sense that discernible random "zero-point fluctuations" in this field remain at the lowest possible energy state. The concept of zero-point energy, particularly as it relates to black body radiation, was introduced by Planck in 1911. Einstein (and others) worked on this too; he called it "residual energy," and his original German term was "Nullpunktenergie" (null-point energy). More recent thinking guides the article by Seife in Science, Vol. 275, 1/10/1997, p. 158, Quantum Mechanics: The Subtle Pull of Emptiness. (A modern review article is available at http://www.calphysics.org/zpe.html.) Even the topic of the widely publicized 2013 Nobel Prize in physics, awarded to Francois Englert and Peter Higgs, supports this concept. They propose a universally pervasive quantum energy field that is responsible for the masses of subatomic particles, including zero rest mass for photons. I.e., via interaction with jiggling particles, this energy confers mass, as in $E=mc^2$. Features of this field may well be a form—or even a consequence—of zero-point energy!

Clearly, zero-point energy can manifest itself in various ways—as can energy in general—including electromagnetic and kinetic. *We can readily envision that jiggling motion arises from random fluctuation in kinetic energy in the zero-point energy field.* In other terms, particles jiggle because they are immersed in a "restless sea" of random energy fluctuations, and they stubbornly persist in jiggling because zero-point energy fluctuations persist stubbornly. (I will pick up these threads in a discussion of an uncertainty principle and in a chapter on fields.)

Experimental evidence for zero-point-energy is found in several ways. For example, a simple—though technically challenging—observation is that liquid helium will not freeze into a solid at near absolute zero, apparently because a hard crystalline state is prevented by residual kinetic energy. We can picture cement that does not set as long as its particles are kept in motion. This phenomenon was discovered over a century ago, though later very cold helium was solidified by adding pressure. (The element helium is inert and very light. These features presumably allow its atoms to jiggle with less restraint than in the case of other elements. [Discussed at length by John Wilks in 1967.] Comparatively vigorous jiggling may prevent the solidification of helium under conditions wherein other elements do harden.)

A more dramatic quantitative demonstration of zero-point energy is the Casimir effect (noted around 1948). Though zero-point energy is very pervasive, it is possible to eliminate some of it from the gap between two very closely spaced smooth metal plates. The plates are set close enough so that only certain electromagnetic particles, notably photons, fit between them. Casimir predicted that the plates would show attraction to each other because of the photons that can exert pressure outside the plates. Indeed, the closer the plates are brought together, the more of the zero-point field between them is reduced, and the stronger their attraction (up to a limit, since the plates are also made of jiggling particles). A simple way to picture this is that many particles (photons) jiggle against the outside of the plates, pushing them together, while few particles are present which can jiggle between the plates to keep them apart.

Subtle evidence for zero-point energy arises in the Lamb shift. We know that the electrons in atoms exist in energy levels called shells, and these electron shells can have sub-shells, each of which has a different shape. When this scheme was discovered—by Bohr and others, representing a huge break-through in science and a triumph of quantum mechanics—the assumption was that two sub-shells of the same shell have the same energy. This energy could be assayed by observing the colored line spectra that arise from energy emitted by atoms.

That is to say, all else being equal, the shape of a sub-shell should not affect its energy, and the line spectra for such sub-shells ought to have the same colors. It turned out (in about 1950) that otherwise identical sub-shells with different shapes do not produce identical line spectra, which means their energies are not equal. This "shift" in spectral lines is attributed to a difference in zero-point energy fluctuations experienced by sub-shells of different shape; it helps to think of the differently shaped sub-shells as intra-atomic Casimir plates. Another more direct but also technically difficult approach to reveal zero-point energy is to trap particles in an energy field so that they cannot move freely. Even such severely constrained particles continue to show some kinetic energy, despite all attempts to suppress it.

My point is that jiggling appears to have a sound physical basis. No particle ever stays completely at rest but is constantly in motion because of a fluctuating zero-point energy field that is constantly interacting with subatomic matter. This means that the basic substructure of the universe is a huge restless quantum field that cannot be eliminated by any known laws of physics. (Here "quantum" means involving very small particles. As I will discuss in my chapter on fields, this concept dominates quantum field theory.)

In the next seven paragraphs I will gradually bring into view the "mystery" I aim to mollify now, the uncertainty principle. We have some indication of the numerical strength of zero-point energy and zero-point energy fluctuations. We will see that in a key equation, Heisenberg's uncertainty principle requires that Planck's constant, h, have a magnitude greater than zero. Perhaps then h is a measure of zero-point fluctuations. Indeed by means of a simple calculation we can estimate the average (but only the average) energy in the zero-point field. This offers a hint as to approximately how much energy exists in jiggling, and it must be very minute.

Planck's constant is also treated as a unit of "action," which is likewise a measure of energy. Apparently, the universe contains a certain minimal amount of energy quantified in h, which means that the universe contains a certain "minimal amount of uncertainty" that expresses itself—in my view—as the randomness of jiggling. Moreover, since h is very small, it follows that jiggling is very delicate and elusive. As mentioned above, thermal vibrations are more evident, notably as heat, and they can be very destructive, whereas jiggling and zero-point energy cannot burn—or even warm—anything. [2]

In the context of energy, I shall raise the following issue: Perhaps the persistent jiggling at absolute zero or in absolute confinement, even if subtle, is not truly random. That is to say, perhaps zero-point conditions do not require zero predictability. Might then the frequency and energy-fluctuation of jiggling be entirely regular and predictable? The answer appears to be "no," and the evidence for this comes from several sources.

---

[2] Susskind, p. 98. See Bibliography starting on page 71.

First of all, in the real world, and even in the laboratory, absolute zero is impossible to achieve. It requires a pure vacuum—absence of all matter, even subatomic particles—which so far has proven to be technically and even theoretically impossible, and which does not appear to exist naturally anywhere in the universe. Likewise it is impossible to trap and fully immobilize particles, even when isolated from all external energy. (Such experiments always yield a range of results; the exact location is never certain.) Therefore, in reality all feasible interactions between particles take place above absolute zero temperature and above truly zero energy.

Second, if particles were cooled to absolute zero and their jiggling did become predictable or even ceased, then the uncertainty principle would be violated. (Jiggling is not only an issue of random location—exactly where a particle is located—but also of momentum—exactly how is it moving.) This is because it would mean that the location of the particles would be known or at least knowable, while their momentum, which relies on energy, would be eliminated. In effect nature protects the uncertainty principle very diligently, particularly with regard to location and momentum, and in my view the basic reason for this is random jiggling.

It may be well now to preview a key point. Heisenberg's uncertainty principle "states" that location and momentum cannot be exactly ascertained at the same time for the same particle (details shortly). In this context my critics accuse me of raising an unsolvable "chicken-or-egg" argument, but I maintain that jiggling is the origin—not the consequence—of the uncertainty principle. Said more fully, the conventional view is that zero-point energy is necessitated by Heisenberg's uncertainty principle, because without jiggling the momentum and position of a particle would both be known precisely and simultaneously. Ergo, in this view particles jiggle so as to obey the uncertainty principle. My view is that *the uncertainty in Heisenberg's uncertainty principle stems from the fundamental randomness inherent in jiggling.* This principle merely demonstrates that no particle is ever at complete rest. In this sense the uncertainty principle arises from and obeys jiggling.

I also think of it this way: If we "freeze" the location of a particle—such as by drastically reducing the temperature—its residual jiggling is represented by random momentum. If we "freeze" the momentum of a particle, its jiggling is represented by random location. In this sense, the reciprocal uncertainty of location and momentum (which is the mathematical hallmark of the uncertainty principle) is preserved, and it is a direct *consequence* of the randomness and persistence of jiggling. This notion brings me to the most direct application of the jiggling postulate in the context of the "mysteries" of quantum mechanics, notably the uncertainty principle: *Jiggling provides a simple explanation of the uncertainty principle.* This is a key concept, so let me now focus on more details.

The very famous and important "Heisenberg [3] Uncertainty Principle," which for brevity I simply call the uncertainty principle, declares that there is a limit on the accuracy of simultaneous measurement of certain observables. ("Observables" are quantities measured or considered in quantum mechanics.) The most notable instance of this principle is our inability to determine the exact position (or location) and the exact motion of a subatomic particle at the same time. In this context, "motion" is better described in terms of momentum (which is mass times velocity; with the customary symbols, $p = mv$, where p represents momentum).

Paraphrasing Heisenberg, "the more precisely the observable for location is determined, the less precisely the observable for momentum is known, and vice versa." In equation form (published first in 1927), this quantum-mechanical principle is now commonly written as

$$(\Delta p)(\Delta x) \geq h,$$

---

[3] Werner Heisenberg was a very productive contributor to quantum mechanics. Besides the uncertainty principle, he is noted for formulating matrix mechanics, a system of mathematics for quantum mechanics.

where the uncertainty in momentum is "(Δp)" and the uncertainty in location is "(Δx)." (These terms are unconventional uses of "Δ" which usually stands for "change.") Whenever Δx increases, Δp decreases, and conversely whenever Δx decreases, Δp must increase; we see that these uncertainties are reciprocal. (The right side of the above equation could be $h/2\pi$, as Planck's constant is easier to use in this form.)

A crucial feature of the above equation is Planck's constant h, which I mentioned earlier; the magnitude of h may be a measure of the average magnitude of jiggling. If h were smaller the uncertainties would be smaller, but they would not vanish unless h vanished, which mathematically implies that the uncertainty principle exists because of a non-zero h. Indeed a key difference between classical and quantum physics is that the former tacitly presumes that Planck's constant is zero, whereas the latter insists that this value (or $(\Delta p)(\Delta x)$) is never zero. In reality, as I pointed out, the value of h is extremely small (about 6.5 x 10^{-27} erg sec), which is why the uncertainty principle is usually evident only in sub-microscopic entities such as subatomic particles. However, in theory size does not matter; under sufficiently detailed scrutiny even large things obey this principle.

There are many ways to express the Heisenberg uncertainty concept verbally. E.g., because h is not zero, whenever the uncertainty of location is high, the uncertainty of momentum is low, and whenever the uncertainty of location is low, the uncertainty of momentum is high. Accuracy in one of these data simultaneously sacrifices accuracy in the other, and we can never know both the simultaneous location and the momentum of a particle—nor, in theory, of any object—with infinite accuracy. We can even aver that the precise determination of the location of a particle precludes—it is utterly incompatible with—the precise determination of its momentum at the same time. If we know how it is moving, we cannot tell where it is, and vice versa.

It is not even essential that we take measurements in the usual sense. For example we say that when a particle has a definite location, it must have indefinite momentum. Therefore, if we could measure a definite location for a particle—though we need not—we would encounter infinitely uncertain momentum for that particle. In theory therefore, uncertainty is inherent in the behavior of particles whether or not they are being measured.

Besides location and momentum, other pairs of measurements—complementary or conjugate pairs of observables—show the same kind of reciprocal uncertainty. One such case involves angular location and angular momentum, though not pertinent here. However very importantly, we cannot determine exact energy in an exact time frame for a particle. In equation form, this version of the uncertainty principle, the "uncertainty of energy vs. time," can be written as

$$(\Delta E)(\Delta t) \geq h.$$

In words, if we wish to determine the energy in a particle exactly, we must measure for a long time. Conversely, a brief measurement (small Δt) yields an inexact assessment of energy (large ΔE). Thus we can quickly approximate strong energy, and we can gradually measure weak energy, but we cannot determine weak energy quickly. (The energy/time uncertainty principle is somewhat different in that time is not strictly speaking a quantum observable like momentum and location; it's a measurable scalar quantity. Nonetheless time is treated like an observable.)

A comment here concerns statements to the effect that greater exactitude in our knowledge of one observable reduces the exactitude in knowing the other observable. This is actuality difficult to apply for location and momentum, since information on location is only as exact as the method of measurement allows. Reciprocal uncertainty is sometimes easier to elicit for energy and time: The more time we allow in measurements of energy, the less variation in the results. Conversely, repeated brief measurements of energy are more apt to vary.

The uncertainty principle dealing with energy and time is paramount in subatomic particle physics, which is underpinned by quantum field theory, QFT; this topic is so important—and also somewhat mysterious—that I devote a separate chapter to it. Suffice it for now that QFT is the most evolved format of quantum mechanics, one that at long last conclusively implicates Einstein's special relativity. The crucial aspect here is the equation $E=mc^2$, expressing the concept that energy and mass are physically interchangeable. One of the key processes (interactions) among elementary particles in contemporary particle physics is the spontaneous and random creation and annihilation of various subatomic particles, apparently "out of nothing."

However, it's not really "out of nothing." QFT holds that various energy fields exist, the most familiar of which is the electromagnetic field (studied in quantum electrodynamics, QED). As is experimentally evident and mathematically predictable, energy can be "borrowed" from this field, and that energy can appear as the mass of a created particle, in compliance with $E=mc^2$. Likewise, elementary particles can be annihilated, and "antiparticles" can participate in such interactions. Familiar examples in QED are the sudden appearance and sudden disappearance of photons, electrons, and anti-electrons called positrons; these processes will reappear in the chapter on fields on page 67.

Here is where the above energy-time uncertainty principle enters the picture: While it is not possible to predict exactly when, where and which individual particle(s) will be created or annihilated—these events are probabilistic—the more energy is "borrowed" to create a particle, the less time the particle can exist. Conversely, less energy may be involved but retained longer, though exactly how much is also unpredictable. Still, more/less energy translates into respectively more/less mass in the new particle. I.e., "lending" more demands repaying it sooner. The implication for my jiggling postulate is obvious: Since energy and mass are interconverted unpredictably, then—though harder to envision—*energy also jiggles.*

Incidentally, what about the laws of conservation or energy and matter in this scenario? On a small scale, when for example an electron and a positron are created together, twice as much energy needs to be "borrowed." On a grander scale, we see how the relativistic mode of these laws forms one intellectually gratifying entity: Physicists of a century ago would describe two principles, "energy in equals energy out," and "matter (mass) in equals "matter (mass) out." Modern physicists, particularly in applying QFT, rely on only one principle, "energy plus mass in equals energy plus mass out." (I discuss this concept as well as the equation $E=p^2c^2+m^2c^4$ at the end of the next paragraph and in my books on Relativity; please see Bibliography, page 71. Conservation will reappear in the chapter "The Mystery of Fields," page 69.)

Also incidentally, here is an example of how QFT implicates, and indeed requires, both quantum mechanics and special relativity; the former contributes the key role of probability, and the latter adds the fact that energy is equivalent to mass. There are other examples: QFT has led to the explanation of a purely quantum phenomenon,[4] quantized particle-spin (known as intrinsic angular momentum), but the full mathematical description of a particle with spin necessitates the equations $E=mc^2$ and the relativistic union of energy and momentum, $E=p^2c^2+m^2c^4$.

Anyway, the uncertainty principles do not mean that we cannot try to measure location and momentum together in one experiment; we certainly can, and likewise for energy and time. The uncertainty principles do mean, however, that in a physics laboratory after we have measured one of these observables, such as momentum, an immediate (or ideally simultaneous) measurement of location is subject to a probability. This may be discernible as a variability of the latter result and a range of results during many runs of an experiment; more on this and the unification of quantum mechanics and relativity under "technical" mysteries on page 61.

---

[4] The QFT equation melding special relativity with quantum mechanics, describing spin, *and* anticipating existence of antiparticles was derived by Paul Dirac in 1928; covered in chapter 20 in my relativity book.

Commonly encountered discussions about the difficulties with measurements are somewhat misleading; the measurements themselves are not the root of problem in situations that reveal the uncertainty principle. While measurements on particles are difficult to obtain, they represent only technical hurdles that can be reduced. Sophisticated experiments do substantiate the uncertainty principle, independently of the technical precision of the experimental apparatus. In other words if infinitely accurate measurements on observables were technically possible, the uncertainty principle would still be evident. (Why? Because jiggling would still be present.)

In the literature of quantum mechanics, we often see explanations of the uncertainty principle for momentum and location by using this kind of analogy: If we seek to determine how a tennis ball moves during a match, we are looking for its velocity of at a location, which is a momentum. We can use a high-speed camera, and the faster the camera's shutter, the better the measurement of location. Indeed on a "stop-action" photograph the tennis ball appears completely still. However, then there is no way of measuring its motion. And if we use a longer exposure so that motion is apparent, the tennis ball becomes a blur with no certain location. Hence the better the depiction of motion, the poorer the measurement of position, and vice versa.

This approach by itself is an inadequate explanation of the uncertainty principle, and it is a variant of the idea that the basic fault lies with our measurements. For example, extremely fast cameras can be devised, and we can even photograph atoms, but uncertainty persists. Indeed these methods merely confirm the impossibility of recording completely accurate positions and motions at the same time. It appears that nothing has concurrent definite position and motion.

Then why is there an uncertainty principle? As I already averred, my short answer is jiggling. But let me invoke a more thorough explanation of uncertainty by asking just how we determine the position of a particle, say, an electron. Of course the answer is that we need a way of visualizing this particle. I.e., we must devise an experiment to see where an electron is located, which means that one way or another *we illuminate the electron.*

But what happens when we shine a light on a particle such as an electron to make it visible? We strike the particle under study with photons! Then ideally the reflected photons should tell us where the electron was. For instance, the photons can rebound into the retinae of our eyes—or first into some kind of camera—whence we can study an image that reveals the particle's position. Let us say that the photons are "illuminating," and that the particle under scrutiny is being "illuminated" and thus is being "seen" so that its position can be measured.

Several points need to be raised about this scenario. First of all, in principle, this is how observations are made, scientific or otherwise; something is illuminated, and the reflected light makes it observable and measurable. Second, the electron under study—the illuminated particle—can be representative of any object of interest in quantum mechanics (or in any branch of science). Third, the illuminating particles need not be photons; they can be other particles, even other electrons (as in an electron microscope), but to keep things simpler let us now consider only photons for illumination. Finally, again in principle, it is possible illuminate only one electron at a time, and we can do so with one illuminating photon at a time.

The problem is easy to envision: Photons are very small, but so are electrons, and when a photon strikes an electron, it will change the electron's location just as we seek to determine that location. Indeed the harder we try to illuminate the electron by using more photons, the more we disturb its location. However, *what if the photon is jiggling? Given the unpredictable nature of jiggling, the exact location of the illuminated electron must also be unpredictable,* or in more apt terms, *its position is uncertain.*

We can restate this concept for momentum. Our illuminating photons possess energy, notably kinetic energy, which means that they have momentum. This in turn means that as a photon strikes an electron, the momentum of the photons is transferred to the electrons. (Please visualize colliding billiard balls.) Meanwhile, no matter what momentum the photon possesses, *by virtue of jiggling this momentum can never be exactly predicted.* Now if the (illuminating) photon can change the (illuminated) electron's momentum, then that photon must have introduced an uncertainty of momentum with respect to that electron. (Imagine playing billiards with trembling [jiggling!] hands. No matter how skilled you are, you can never be sure which way and how fast a ball will recoil when the cue ball shrikes it.) Again it appears that this particle has no concurrent definite position and motion. These "definites" simply do not exist!

This approach does not address the following objection. Since a photon is one quantum of energy [5] that corresponds to a specific amount of momentum and even some known velocity, why not assume that an illuminated electron always absorbs the entire quantum of light energy? In that case perhaps we could predict the disturbance introduced by the incoming photon on the reaction of the electron under study, so that the observed momentum of the latter would not be uncertain. In other words, by virtue of quantization the characteristics of quanta may be calculable, so that perhaps we can allow for their effect and thus circumvent all uncertainty.

In reply we invoke an important process in subatomic physics: When illuminated, the electron absorbs the quantized energy of the incoming photon. This electron then emits another photon with less energy. *However, the emitted amount of energy is subject to a probability; if nothing else, photons jiggle, not to mention electrons.* The observed electron is free (not quantized) and can absorb any amount of energy from zero up to the energy of the incoming photon. The difference between the energy of the incoming photon and the energy of the emitted photon is *unknown*, and it is retained by the electron in the form of a new but uncertain momentum.

We see that in principle, with every observation there is some probability for the amount of absorbed energy—and hence for the observed object's new measured velocity and momentum—and there is no certainty. In short, any act of observing must introduce uncertainty *because the process is unpredictable*, so that even the best the experiment cannot reveal the exact location and momentum at the same time. Aha! Heisenberg's uncertainty principle.

In his book "The Dancing Wu Li Masters" (page 112) Gary Zukav writes "*...we cannot observe something without changing it.*" [6] (The italics are Zukav's.) In my view, even this statement does not go far enough. We know that observation induces an irreducible change, but the gist of the issue is that we cannot predict how much change. Fred Alan Wolf in his book, "Taking The Quantum Leap" (page 80) [7] says this explicitly: "Light act[s] in a disruptive manner whenever it interact[s] with matter."

I also prefer to restate this concept forcefully: Illumination disrupts observation, but by virtue of jiggling, we cannot predict exactly by how much. Hence the critical feature of the process of measurement is an unavoidable limit on how much knowledge we can glean from that measurement because we never know its exact effect. This issue is part of the "measurement problem" of quantum mechanics; measurements are unpredictably obtrusive. Another related thorny question is just what constitutes an "observation;" does the act of looking preclude obtaining a true measurement; is the "looker" always a part of the outcome? (A famous mathematical counterpart of this issue in quantum mechanics is exactly when and how does Schrödinger's wave equation "collapse," but this controversy is not pertinent here.)

---

[5] I will cover quanta and quantization in a separate section.
[6] See Bibliography starting on page 71.
[7] See Bibliography starting on page 71.

Anyway, we can also see this process from the photon's vantage, still assuming it is the illuminating agent. When such a photon strikes an observed particle, it gives up some of its momentum, but because it is jiggling we do not know how much. Either way, the reflected photon does not have the information we seek. Said simply, a photon must bump into an electron to tell us anything, but we cannot determine the full effect of the bump. But how can this be if we say that a photon has no mass? To be precise, a photon has no *rest mass*, but it is never at rest (explained in my book on relativity. Bibliography page 71.) The statement that photons move more slowly through a medium is inaccurate; they have less macroscopic linear speed, but considering jiggling, their microscopic speed remains "at the speed of light, c."

To summarize this crucial concept let me use an expanded analogy, based on Heisenberg's reasoning, one that emphasizes and illustrates the very root of the uncertainty. In order play well, a skilled tennis player needs to know exactly where the tennis ball is located. To visualize the ball, photons are needed, which are reflected from the ball to the player's eyes. Of course the photons which strike an object as large as a tennis ball hardly make a difference, even as they jiggle. But if the player were aiming for a molecule rather than a tennis ball, the jiggling photons striking the molecule must displace it, and the amount of displacement is both significant and unpredictable. If the player made many observations of the location of the molecule, the individual results might be clustered around one result, but they would still never be all identical, and the variance would never be fully predictable. The observant player would always be somewhat unsure, and his or her game would never be perfect.

Meanwhile a group of tennis enthusiasts is watching. One naive fan points out that fewer photons would be less disruptive. ("Play at night, and turn down the lights.") However it turns out that with too few illuminating photons, the molecule becomes invisible. ("But I can't see the ball [or molecule] in the dark!") That is to say, the player needs enough light to discern reality. And then the few remaining photons still jiggle, so the problem might even be compounded.

Another more sophisticated fan argues that perhaps the player can examine every illuminating photon after its contact with the illuminated molecule. Then such examinations might reveal how much energy each illuminating photon lost and hence how much the illuminated molecule gained. However this would require *another randomly disruptive illuminating photon*, and the same kind of uncertainty would prevail. One way or another, the act of measuring would alter what is being measured unpredictably, always reducing the reliability of the measurement.

Finally an inventive fan suggests using photons that are infinitely small and weak so that they do not disturb the molecule, i.e. the illuminated molecule is comparatively huge. However, a search for such photons reveals an astounding fact, one that lies at the core of modern science: no infinitely small or weak photons exist! That is to say, light is quantized. Worse yet, no photons, even if they could be smaller, would be free of all energy; light always jiggles. The same would apply if a different illuminating particle were used, such as an electron; jiggling still precludes unlimited accuracy and ensures at least some uncertainty, and no amount of improvement in equipment or athletic skill will allow flawless tennis.

We could even argue that were it not for quantum uncertainties, the future might be predestined and predictable. I.e., if all initial conditions were known, and if we applied enough mathematics, couldn't all outcomes be forecast? Then any player could always play perfect tennis and predict the final score! The quantum reply is that this classical concept of physical cause and effect is incorrect. In terms I used earlier, in our universe nothing possesses concurrent precise and unequivocal position and momentum. Such data simply do not exist.

Ergo, in keeping with our current analogy, an incoming photon does not have a unique effect on the electron under study. By its jiggling, the photon only has an unpredictable range of possible effects. The best we can do is to calculate the width of the range, which is what Heisenberg's uncertainty equations (and Schrödinger's wave equations, as I will discuss shortly) can accomplish. Probability remains a fundamental feature of physical nature, evident in quantum mechanics. This feature also is notably absent in classical ("Newtonian") physics and even in relativity; these sciences are said to be "deterministic:" In contrast, quantum mechanics is not thusly deterministic; all it can "determine" is a probability.

Let me carry this thought further in three directions: First, if it were not for the randomness in jiggling, could all of the past be reconstruct-able and all of the future predictable? Conversely does the random, fragile and universal nature of jiggling explain why everything is not the same anywhere? For instance why is the universe "lumpy?" I can argue that if it were not for the jiggling of the miniscule parts of the primordial universe (just after the "big bang" during cosmic "inflation"), every planet—and indeed all locations in the universe—would be identical.

Second, again given the randomness and fragility of jiggling, does nature thus permit—or even require—the existence of a deity? A modern issue in this context is whether a scientist can believe in an effective God (and accept divine miracles). Here jiggling may provide an escape from this dilemma. Walking on water is extremely unlikely but not physically impossible.

Third, if you prefer a wider definition of religion, does jiggling imply that we are all interconnected and participate in a universal consciousness? In other words because of the fragile but universal nature jiggling, each of our actions, and even each thought, indeed may reverberate throughout the universe.

In any case, the concept of fundamental uncertainty is the underlying reason quantum mechanics never yields 100% certain results but only probabilities—an idea that Einstein could not accept, claiming that "God doesn't play dice with the world." (See W. Hermanns, Einstein and the Poet, Brandon Press, 1983: p. 58, and other biographies.) This is also why quantum processes seem to be irreversible; we cannot study the outcome of an event (e.g., where and how the tennis ball bounced) and thereby reconstruct the initial conditions (precisely how the ball was hit with a racquet), because some probability always affected the processes in between. I.e., that process is not deterministic. A corollary of this idea is the conclusion that the flow of time is a unique one-way "arrow" (and hence bars backwards "time travel.")

These concepts are reminiscent of *chaos*—in the physical and mathematical sense—which has attracted much attention; e.g. see Gribbin, pages 73-76 (Bibliography on page 71). One intriguing issue is the relationship between quantum randomness and the randomness that appears in chaos. An interesting question is whether quantum mechanics is the only source of inevitably random behavior in physical systems. I now will show that despite superficial resemblances, quantum mechanics and chaos theory are quite different, particularly regarding the roles of initial conditions (the detailed physical circumstances at the start of an event).

Chaos arises *when seemingly negligible initial differences in dynamical systems are magnified and lead to unpredictable "chaotic" differences in outcomes*. This explains the famous "butterfly effect:" A beat of its wings may cause a powerful storm far away. Quantum randomness—as in the uncertainty principle—implies that *initial conditions can never be sufficiently known*. The elements that constitute an initial condition must jiggle, and *this* is why the final condition, magnified in a system or not, must be subject to at least some unpredictability. On short, our universe is not deterministic; it is ultimately probabilistic.

Let me digress into more detail: It is useful to explore the origin of unpredictability in "chaotic" systems vs. in quantum systems. Another example is easier to dissect. Freeway commuters are familiar with disruptive traffic jams that—surprisingly—are not caused by obstacles such as accidents, stalls, construction, etc. After passing through a jammed stretch, drivers cannot understand why traffic had stopped, except that at some point a few cars may have slowed slightly, and yet this trivial event initiated a gradually worsening process culminating in the jam. That process is an instance of chaos theory.

According to this highly mathematical branch of science, it is quite natural that the slowing of one car in a line of cars constitutes an "initial condition" in a "dynamical system" that evolves over a period of time (which can be estimated) into the slowing or stopping of hundreds of cars miles away. A very slightly different initial scenario—a somewhat different "perturbation" such as a usually negligible difference in the slowing of one car—can unexpectedly develop into an even more dramatic event such as a multi-car pileup, or it can evaporate into insignificance with no disruption of traffic. Such sets of widely divergent outcomes are "chaotic." The salient feature of these seemingly erratic events is *a mathematically exponential hypersensitivity of processes to small changes in initial conditions.* (Exponential values can inflate solutions nonlinearly.) The process can be deterministic (please see page 13) but its outcomes appear to be random and astonishing; they are probabilistic.

The mathematical details of chaos theory are very complex. Three factors are usually cited as prerequisites for chaotic behavior of a dynamical system. However close analysis suggests that the first factor is a consequence of the second and third, and even the relative primacy of the latter two factors is debatable. The three factors, listed in one of the traditional orders, are:

1. Sensitivity (hypersensitivity) to initial conditions, as exemplified above.
2. Topologic mixing, as explained below.
3. Density of periodic orbits, as explained below.

Here I interject the following concept: Initial conditions can be complicated and subtle (e.g. drivers vary in reaction time and driving habits, cars' brakes do not all work the same, conditions vary with the weather, etc.) but my point is that *random quantum jiggling need not be invoked to account for "chaotic" outcomes.* For instance it is extremely unlikely that the jiggling of subatomic particles alone can explain erratic traffic jams or misleading weather forecasts. However chaos theory alone suffices elegantly; initial randomness is magnified, and the process is exponential. (This is why local weather forecasting beyond about a week is so difficult.)

Now please note the above three factors. The first, high sensitivity, can be summarized in an equation containing two exponents; one quantifies the sensitivity to initial conditions, and the other approximates the time required for chaotic behavior to appear. For the two last factors I offer only loose analogies to avoid the difficult math. "Topologic mixing" roughly implies that more than one car must be involved and that traffic consists of a mixture of vehicles on the same road. (In the case of weather, there is a "mixing" of winds.) "Density of periodic orbits" roughly implies that the cars are closely spaced but not sitting on top of each other, all the while moving in approximately the same way.

Mathematically the behavior of chaotic dynamical systems is studied in "phase space" whose dimensions are position and momentum, but not all collections of cars, and no solitary car, will exhibit chaos. In any case, the nature of quantum uncertainty is incidental to outcomes that are chaotic due to hypersensitivity to initial states, and chaos theory alone can account for *the apparent unpredictability in such outcomes.* Please note the subtle distinction: Quantum jiggling is a property of initial conditions that injects basic randomness evident in small-scale events. Chaotic hypersensitivity to small changes in initial conditions—for whatever reasons—injects growing randomness into some large-scale events that exhibit the above factors.

So much for chaos; back to quantum physics and its math: We recall that the h in Heisenberg's equation (page 7) is an irreducible number, minuscule but never zero. Here I reiterate that jiggling is inescapable. I again state that h provides an estimate for the least vigorous jiggling that is possible, and it too is never zero. Moreover, the probabilistic aspect of jiggling is likewise never zero, and hence I reiterate: Probability is a fundamental property in physical nature.

I even consider it possible to "derive" Heisenberg's equation non-mathematically from the features of jiggling. Please imagine confining a particle until it can hardly move, thereby minimizing its possible locations so that Δx *is reduced.* However given the persistence of jiggling, the particle "resists" its confinement vigorously and with greater momentum, so that Δp *is increased.* The reverse is also easy to picture: hindering its momentum by some means forces the particle to jiggle more widely. However there is a limit to either reduction, and that limit is Planck's constant h. (Various formal mathematical proofs of Heisenberg's equations do exist, and at least one of these appears in almost any detailed text on theoretical physics.)

Surely by now my message is clear. *Heisenberg's uncertainty principle, critically important in subatomic particle physics and notable in popular literature, need not be a mystery at all. At its root lies jiggling, and the "uncertainty" in this principle arises from the unpredictability of jiggling.* In other words the uncertainty principle merely represents a prominent manifestation and a consequence of jiggling. In my next section the uncertainty principle will boldly resurface as the basis of another notorious mystery of quantum mechanics, wave-particle duality.

But first let me place my concept of jiggling into a wider picture by focusing on the notion that this jiggling is *motion.* In my view it is a crucially important form of motion, central to quantum mechanics, but indeed motion in general is the reason physics exists. Please think about this: We engage in physics to understand the motions of galaxies, stars, planets, parts of machines, projectiles, electrons, and other elementary subatomic particles. In every case, to understand how things move, we use mathematics to devise equations of motion.

A very basic "equation of motion" of course is Newton's seventeenth-century F=ma, and ever since then generations of scientists have found more sophisticated and accurate equations: To mention four pertinent ones, we have Einstein's field equations of general relativity for gravitational motion, Schrödinger's wave equations for quantum systems, Dirac's equations incorporating special relativity and quantum mechanics for electrons, and Feynman's path integral equations for why objects of all sizes move as they do. In this context all I have done is focus on one form of motion, jiggling, I describe its features, and I use these features to help me, and hopefully to help you, understand the alleged mysteries in one branch of modern physics.

Wave-particle Duality and Waves of Probability

Here I first introduce some history, and later I will invoke THE WAVES-OF-PROBABILITY POSTULATE to explain away the apparently formidable mystery of wave-particle duality. The standard definition of wave-particle duality is the principle of quantum mechanics that *subatomic particles sometimes show particle-like behavior and at other times show wave-like behavior*. This idea sprang from the observation attributed to Louis-Victor de Broglie (in 1923-4) that a certain equation contains a variable for waves (wavelength, λ) and a variable for particles (momentum, *p*). This equation, written in its standard form

$$p = h\,/\,\lambda,$$

was interpreted as indication that one entity, such as a particle, has a wave property, wavelength, as well as a particle property, momentum. Historically, the study of these physical alternatives—waves vs. particles—is much older. Huygens (1629-1695) proposed a wave theory for light, while Newton (1642-1727) favored a particle- ("corpuscular") explanation. We can say that in response to de Broglie's mathematical discovery in the 1920's, quantum mechanics began to teach that both explanations are correct; hence the term "duality."

In my view, the implications of de Broglie's equation are overstated. The wavelength (λ) in the derivation of the above equation is equivalent to the λ in the Planck-Einstein equation

$$E = hf$$

for the energy in a particle (here a photon), inspired by Planck's concept of quantized energy. Here *f* is frequency, which is inversely proportional to wavelength, λ, as longer waves have lower frequencies. In this context λ or *f* is a measure of the particle's energy (more precisely, its quantum energy field), and hence de Broglie ascribed wave- and particle-natures to photons. Waves and particles both have energy, but a wave-particle duality is not essential to this concept. De Broglie proposed other quantum-mechanical theories, some of which have been abandoned, such as that particles are guided by "pilot waves." (The rarely used terms "wave packets" and "matter waves" are examples of waves of probability, covered later in this chapter.)

Incidentally, frequencies and wavelengths of electromagnetic radiation arise prominently in the study of color, in the "tuning" of devices such as radios, and in the biological effects of radiation. For example, violet light has a higher frequency or shorter wavelength and more energy than red light; hence the terms ultraviolet and infrared, and hence ultraviolet radiation is biologically more harmful than infrared radiation. An "atomic" explosion, which is obviously very energetic, releases electromagnetic radiation with extremely high frequency or very short wavelength. In any case, de Broglie's wavelength is an indication of a particle's energy, so of course wavelength can be calculated from momentum via $p = h\,/\,\lambda$; momentum reflects kinetic energy.

Of course frequency and wavelength conjure up images of waves, but I re-emphasize two points: In this context they are measures and properties of energy, and the waves in this context are not exclusively diagnostic of wave-particle duality. In theory, wave-particle duality does not hinge on the amount of energy exerted in experiments, though such experiments are often easier with strong visible light. In short, particles can have energy without being waves.

A significant point about wave-particle duality is how it has been expressed and envisioned in the context of quantum physics: This principle is traditionally said to imply that *one object*, which can be a subatomic particle or something much larger, *can possess two seemingly incompatible natures, a wave nature and a particle nature.* Worse yet, it appears that such an entity can switch back and forth from one manifestation to another, depending on how it is observed, but that it will not show both natures at the same time.

Indeed analogies about wave-particle duality appear in the literature which are downright misleading. For example, we are told that subatomic particles can switch their nature "like chameleons switch colors," or they have "split personalities," or particles have "two sides like a coin," or they act like "Doctor Jekyll and Mr. Hyde." The obvious problem is how something can be a wave (perhaps undulating up and down like the ocean) and at other times a particle, definable as a solitary object (perhaps moving like a bullet).

In the scientific community opinions hold that wave-particle duality is a deep mystery. As I will cover in some detail later, the critical observations about wave-particle duality are made on photons in "double-slit" or "two-slit" experiments. The pertinent issue is that such experiments reveal several results that are surprising and difficult to explain. For instance, under proper conditions one photon appears to pass through two slits or it appears to "interfere with itself." Indeed some scientists consider the latter finding to be a "central mystery" of quantum mechanics. Richard Feynman felt this way but then corrected himself, saying that in fact it is the "only mystery" of quantum mechanics, and interference is a key element of this mystery. [8]

Clearly wave-particle duality constitutes a celebrated and persistent enigma in modern science, justifying my detailed attention. After years of research and dozens of ingenious experiments, many theoretical physicists (e.g. Feynman) have concluded that we simply must accept these observations as facts of nature that cannot be explained, derived, deduced, or predicted. Nonetheless I posit that by accepting comparatively simple concepts, this concession is unnecessarily pessimistic. My POSTULATE for grasping wave-particle duality in this context is

The waves in wave-particle duality are waves of probability.

Here I like to use an analogy. A city police department has devised a map to show criminal activity. A dot on this map indicates where a criminal has struck. One day new data showed a rise in the crime rate in parts of the city, and this was revealed on the map as more dots in these parts. I.e., the dots appeared even more densely spaced in some areas, reflecting that certain locations have become more dangerous. After a year-long "crack-down," a revised map showed fewer dots in those areas. The mayor then declared that there had been a *"crime wave."*

The point of course is that the "wave" depicted on the map reflected a greater probability of criminal activity at some locations. It did not mean that criminals had turned into waves, nor that crimes or criminals had attained the characteristics of waves. Each perpetrator was still one individual, and each crime was still just an ordinary (albeit illicit) event. However, only the pattern of the criminals' behavior in terms of their locations suggested a wave: On a diagram, the number of dots in some places merely swelled and then ebbed in a wave-like manner.

To dramatize this event, a particularly dangerous location in the city was selected and a graph was devised to show how the crime rate increased, crested and then declined. Not surprisingly, *the graph showed a wave*, which was recognized as a wave of criminal probability! In other words the probability of finding someone breaking the law at certain locations is such that when viewed in an appropriate

---

[8] Feynman, p. 1-1 and 1-9. See Bibliography on page 71. Feynman also authored a novel mathematical system of quantum mechanics which showed why the paths of particles ("histories") have different probabilities, and why certain "histories" are most likely to be observed. I cover this from page 57 on.

manner, wave-like aggregates are noticeable. Of course some crimes were worse—perhaps more "intense"—than others, but such variation is not pertinent.

Now I will revisit my crucial point in this analogy: Did the criminals behave like waves? Did they manifest a "wave nature?" No. All that happened was that during some interval of time criminals perpetrated more crimes in some places and fewer elsewhere; *what was "wavy" was the probability of crime;* what generated the wave was a change in geographic density of criminality; what appeared wave-like was only the statistical distribution of criminal activity; what was witnessed was *a wave of probability.*

The lesson in this analogy is simple: Things studied in quantum mechanics, especially particles, do not undergo mutations into or out of wave-like entities. Nor need they be envisioned as moving in wave-like manner. Particles should not be imagined as turning into waves, nor v.v. When the motion of particles can be measured, they are like particles, not like waves. These particles may have energy (intensity) which can be quantified by wavelengths and frequencies, but that is optional and not pertinent here. What is pertinent is that *the probability of finding these particles at some locations is wave-like. The waves in wave-particle duality are not made of wavy particles; they are made of the probability of finding the same kinds of particles at various locations.* As in my analogy, they are waves of likely positions.

Yes, subatomic particles are unlike ordinary objects (e.g., one particle can emit another), but I reiterate that they are still particles; their "wave nature" resides in their probabilistic behavior as revealed when we plot their positions. Thus if we ask, "what is 'waving' in wave-particle duality," the reply is "the probability at locations." Said more forcefully, particles really have no wave-nature or wave-properties, but if we make a diagram or graph to show where particles are likely to be, that picture may reveal patterns that look like waves.

We should note that the above-described wave-like behavior can be revealed in other terms besides locations. E.g., measurements of momentum or energy (see page 22) can elicit wave-like fluctuations, as these observables are also subject to probabilities. However probability and wave-like behavior are more prominent with respect to location (or position), and scientific studies on wave-particle duality have typically focused on locations of the particles. Please consider a metaphor: Waves of probability do their waving "on a sea of all possible results of measurements" (though "in a space of all possible results" is more precise).

Let me also mention some pertinent—albeit specialized—quantum math, because a certain concept puts a mathematical spin on my views about wave-particle duality. A wavefunction, a.k.a. a wave or state function, is a mathematical device (derived by Schrödinger) that encodes the complete quantum description of a quantum system, from which the probabilities—and *only* the probabilities—of all measurement outcomes can be calculated (by means of "operators."). A wavefunction can always be written as a superposition of quantum states, each characterized by a (mathematically complex) probability amplitude, a possibly negative coefficient; the square of that coefficient gives the probability for that state to appear on measurement. In this context, wave-interference is interference between probability amplitudes of waves, and hence this wavefunction is said to describe a "probability wave." Here is a wavefunction $|\Psi>$ written in standard vector form, describing the position of a particle in two dimensions, these labeled x and y. Of course the particle can be a photon, and its position can be recorded on an x-y graph.

$$|\Psi> = 1/\sqrt{2}(\alpha|x> + \beta|y>)$$

(The symbol Ψ can stand alone to denote a wavefunction, or it can refer to the entire equation. This kind of equation is a superposition.) Here the α is the probability amplitude that this particle will be found at location x, and α^2 *is the probability that this particle will be found at location* x. Likewise, the β is the probability amplitude that this particle will be found at location y, and β^2 *is the probability that this particle will be found at location* y. Conversely, α and β are the square roots of the probabilities for the results of measurements. For example, $P_x = \alpha^2$ and $\sqrt{P_x} = \alpha$. That is the reason "$1/\sqrt{2}$" appears in the above equation. (Here the 2 implies a 50-50 chance. As there are two specific locations in this case, we say that two "eigenfunctions" exist.)

When a graph is plotted in the case of interference, waves may appear in the graph. But—and here is my current point—when interference is evident, *it is because the probability amplitudes in the above equation fluctuated in a wave-like manner*. The particle does *not* form a wave; in fact the particle does not even appear in the equation. Only the probabilities of its locations (including points on a mathematical field) form waves, calculated as squares of its probability amplitudes. Thus, from a mathematical vantage, these waves again are *waves of probability*. Earlier I showed de Broglie's equation which suggested—albeit misleadingly—that a mysterious wave-particle duality exists. That notion was fed by ambiguity as to exactly what is "wavy" in Schrödinger's wave equations, until Max Born concluded that *their solutions describe where particles with a given mass will probably be observed* (via the square of Ψ).

Now I turn to the type of *experiment* that elicits what is called wave-particle duality. Indeed, as I will stress later, wave-particle duality is an interpretation of a particular experimental result, and wave-particle duality owes its renown to our ability to perform that kind of experiment. Here is a diagram of the typical setup. A source of light is used, and Einstein's work with the photoelectric effect already indicated that light consists of particles (photons). The light—the beam of photons—passes through two vertical narrow parallel slits in an otherwise opaque barrier. The key issue is, what will appear behind the barrier on the screen where the "?" is placed.

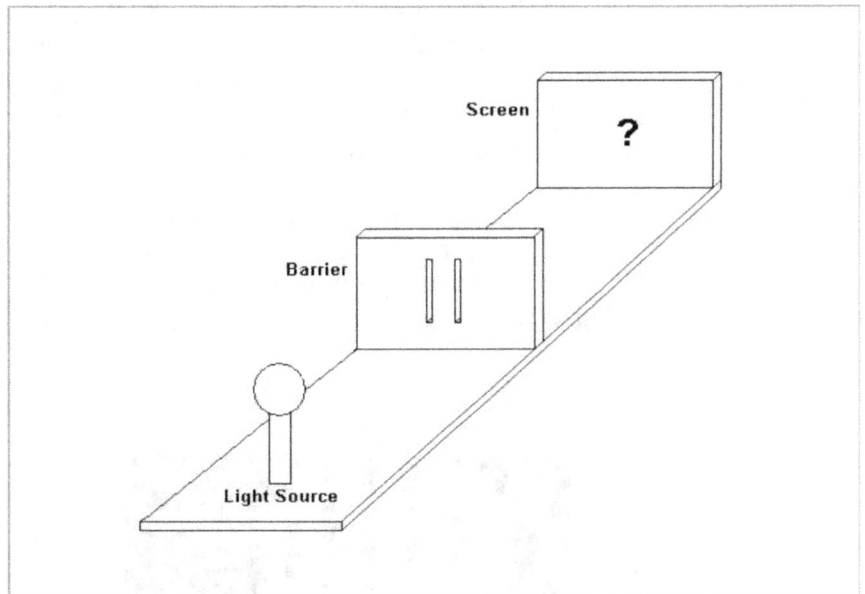

The initial common-sense expectation—and the reasoning via classical physics—is that we will see two vertical bands of light on the screen. However, the crucial finding in such an experiment with small particles is that *an interference pattern is projected onto a screen set behind the barrier in line with the light source, and *this pattern is a series of many alternating bright and dark vertical bands*. Such a pattern can only be explained by wave activity.*

The standard interpretation, which I am challenging here, is that because we can find interference in a setup where photons are involved, light (photons) must have a particle-nature *as well as a wave nature*. Ergo we are told that the presence of an interference pattern necessarily indicates the existence of the enigmatic "wave-particle duality." (Befitting modern terminology, I also suggest that a better term is "wave-particle complementarity," a concept derived by Bohr. Historically Bohr's thinking and writing are somewhat murky, but in brief his contention was that complete quantum descriptions of things may require two possibly clashing facets which complement each other, such as a wave- and a particle-nature.)

In the next diagram we see such an interference pattern; each dot is where a photon landed on a screen. Please note that in this case five bands with fringes are seen, not two, and that number may be higher than five if the apparatus is wider and more sensitive. Even this picture, with no further manipulation, suggests five parallel waves, and even with a minimal knowledge of physics we recognize a typical pattern of wave interference. Now I grant that what is visible on the screen, specifically interference, is diagnostic of waves, but I maintain that nevertheless *no exotic dual nature needs to be invoked*, and that no mystery needs to be feared.

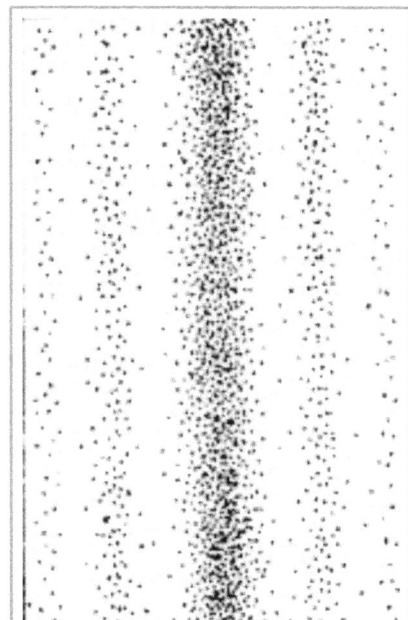

Since each dot is where at least one particle landed, we can also say that we are seeing waves of *events*. This notion coincides with my "crime wave" analogy (each crime is an event) but the more intuitive concept is that each dot is a location where a photon landed on the screen.

Incidentally, we note that the dots vary somewhat in size and intensity. We can surmise that this variation stems from a variation in the energy of each photon as it encountered the surface. This energy, in appropriate units (e.g. joules), is the E in the equation $E = hf$. (Given also de Broglie's $p = h/\lambda$ [p is momentum], the energy E is kinetic, so that if the photons were sufficiently energetic, we might see holes with burnt edges in the screen, and a wavy pattern would appear nonetheless.) I reemphasize that although a frequency (the f) appears in the equation, and although frequency is a general feature of waves, the diagram shows the wavy distribution of individual particles independently of these facts.

As photographic film is typically used as the screen to record the findings, the image is usually a negative (the dots are white or bright but they usually run together, forming fringes), which is also more intuitive than black dots. One such pattern as it is often seen in real experiments is this, where P_2 is a dark band and P_1 is a bright band:

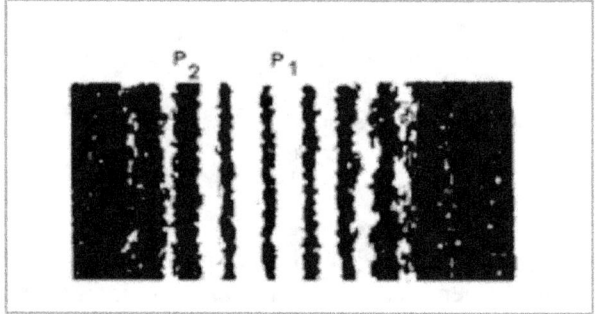

A bright band is where many photons landed, and a dark band is where few photons landed. I will come back to the irregularity of the edges (fringes) of the bands, as this detail is significant, but I stress that the pattern arose from a collection of dots, each representing a particle.

Now as is critically important, when such an interference pattern is converted to a graph, the typical result is a series of *waves*! The next diagram shows a wider interference pattern and, just below it, is its graphic representation. The latter clearly shows a wave, with higher peaks (greater amplitudes) toward the middle. As I said and will discuss further, the tempting inference is that uncanny wave-particle duality has been demonstrated, but this conclusion is misleading.

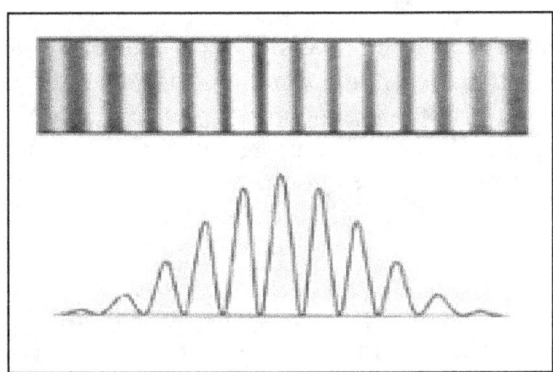

I will also elaborate later on what else is done in the performance of such experiments, but first I must dwell on a key point: It is possible to release photons form their source one at a time, and it is possible to allow each successive photon to leave its localized mark on the screen. Of course this is just what is expected of particle-like photons: We can record the final location of each photon one-by-one. As the experiment progresses with more and more photons, it is obvious that the interference pattern is gradually generated *because photons are more likely to land at some locations and less likely to do so at other locations*. In other words the observed wave-like behavior—the interference pattern—is the consequence of the varying probability of photons (again, particles!) landing at the various locations. Therefore, the alternating light and dark fringed bands forming the pattern represent *interference of waves of probability*.

To my mind nothing mysterious happens when evidence for "wave-particle duality" is observed as above or as in typical wave-particle experiments. Photons are and remain particles whose arrivals are simply clustered as events on the screen at locations of high probability, and they are rare at locations of low probability. *When these contrasts are presented in an appropriate manner, typically as a graph, a wavy pattern with fringes may appear.* As for de Broglie (page 16), his wavelengths were "waves of matter" and its energy; they were not waves of the probability of finding matter (particles) at various locations.

One of the celebrated developments in experiments on wave-particle duality is the use of electrons and x-rays rather than photons. Though technically more difficult, interference patterns can still be observed, and the traditional interpretation is that wave-particle duality is a common phenomenon not restricted to visible light. However I still envision the landing of particles in wave-like distributions on screens, and the particles have not changed their nature. What we see are still basically waves of probability.

An important detail is why photons in general (not just in two-slit experiments) arrive in clusters with fuzzy and irregular edges rather than in sharply demarcated zones. My answer of course is, *because they jiggle*. In fact I can say that the appearance of an interference pattern as obtained in actual experiments— with fuzzy and irregular but discernible bands—is attributable to *two* processes: the formation of waves of probability where photons land, and their jiggling while they were en route. In this paragraph the emphasis is on the former process.

Another notable point is that interference is a feature of wave behavior in general, and *it is not unique to quantum physics or to two-slit experiments*. Interference occurs and can be observed under many circumstances, natural and experimental, whose common element is various waves meeting in space.

Even waves of fluctuating energy can interfere (mentioned on page 18). [9] Whether these waves have anything to do with quantum mechanics is immaterial to the appearance of interference. Nevertheless here I will restrict the discussion to light and photons in the context of quantum theory, and the message I wish to convey is this: *The evidence for light apparently behaving as a wave or as a particle was gleaned from the results in two-slit experiments that actually reveal interference between waves of probability.*

Further clarifications are needed regarding two-slit experiments. The above-diagrammed two-slit experiment is but one example—though a prime example—of a class of experiments aimed at revealing the wave-behavior of particles. In fact the groundbreaking experiment by Young (around 1800) did not even have two slits; he just split a sunbeam in two using a thin card. Modern versions use crystals to split beams, and as I just mentioned, the beams need not be made of visible light; they can be x-rays, electron beams, etc. Still, it is customary to call all such arrangements "two-slit" experiments, and to think of them in terms of visible light.

In practice, the light-sources for successful two-slit experiments are difficult to design. (Young was very ingenious and somewhat lucky.) The best results—the clearest interference pattern of many bright and dark bands—require that the two beams emerging from the slits be made of coherent light. Such light has all its waves in phase, and it consists of only one wavelength (which means it has only one color [hence also one frequency]; it is monochromatic, thus also avoiding color dispersion. To visualize a beam of coherent light, imagine a long parade of well-rehearsed soldiers marching in step). A sophisticated source of coherent light is a laser, but one shortcut is to use a steady source of monochromatic light which is exactly equidistant from the two slits. The equal distances ensure that the light remain sufficiently coherent, as discussed in the next paragraph and shown in the next diagram.

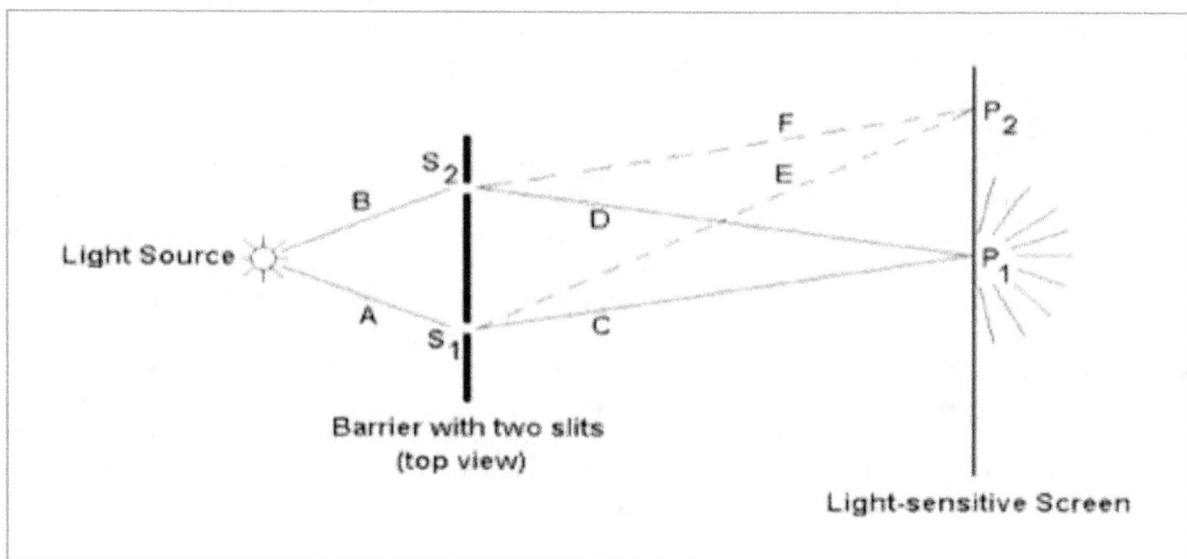

The above diagram serves as a more schematic overview of a typical two-slit experiment, but it will help make a particular point. Please imagine we are looking down on the setup. We see a light source on the left, which produces rays of monochromatic visible light. The two slits in the barrier now appear as gaps, labeled S_1 and S_2. Rays A and B are exactly the same length, so that A reaches slit S_1 exactly when B reaches slit S_2. This means that each slit acts as a source of sufficiently coherent light, and the rays are in phase as they leave the slits and move toward the light-sensitive screen on the right.

---

[9] See reference (in Bibliography starting on page 71) to the 2015 paper by L Piazza et al, where *quantization*, which is particle behavior, and an *interference pattern*, which is wave behavior, are *observed simultaneously*. The particles here are electrons, and the waves are waves of energy. The authors do not stress that the latter are influenced by probability, which would make them waves of probability as I posit.

Here I will insert an important semantic detail we should be aware: The interaction between two beams in a two-slit experiment is called "interference" even though the beams can add to each other—in "constructive interference" which produces bright bands—and cancel each other—in "destructive interference" which produces dark bands—while both occur at the same time. Therefore the bright bands are the consequence of wave interference, even if this terminology is not intuitive, and the same words can be used for any case of wave interaction.

Let us consider rays C and D first, which remain in phase with each other all the way to point P_1 on the screen; i.e. pairs of photons arrive there together, since the length of C equals that of D. This means their brightness is additive—they show constructive interference—at the P_1 location. In three dimensions, we would see a bright line passing through point P_1, which on a graph would look like part of a wave. Again, I stress that this wave represent the probability of photons landing at location P_1, and the fringes reflect the (fuzzy) nature of this probability.

Next let us consider rays E and F, noting that E is somewhat longer than F, simply because the distance from S_1 to P_2 is greater than from S_2 to P_2. Since the waves in E and F are no longer in phase with each other when they reach the screen—photons in ray E had longer journeys—they cancel each other, so that point P_2 is dim. In other words crests from two waves do not arrive at the screen together. (We must assume that this difference in length is not exactly one wavelength, or else the effect we wish to show is negated as the waves are in phase again.) In three dimensions, we would see a dark band passing through P_2, and on a graph we would see the gap between waves. This means that the waves of probability in rays E and F created destructive interference at this location. Thus alternating crests *and* troughs are generated. *The pertinent lesson here is that this interference pattern is basically an effect of two waves that meet under such circumstances. In a quantum-mechanical experiment these are waves of probability, but the patterns we see are still explained by simple physics and geometry.*

Again I stress that these waves need not have any connection with quantum mechanics, nor must they be made of photons, nor are they found only in laboratories. For example in the above diagram, the "light source" could be a source of water, the "barrier" could be a breakwater in a harbor, and the "slits" could be gaps in the breakwater. If the water surface is calm enough and appropriate observations are made, *a wavy interference pattern will appear.* Arial photographs of harbors readily show this, entailing nothing mysterious or dual about the nature of water. No individual water molecule (each a particle!) had to acquire a wave-nature.

The above diagram would become needlessly cluttered (with pairs of rays extending to points P_3, P_4, P_5, etc.) but as we saw, a well-designed two-slit experiment using visible light shows a series of bright and dark fringed bands around the center of the screen. To re-emphasize the current theme, *such an interference pattern is diagnostic of wave activity.* We may also again invoke Einstein's explanation of the photoelectric effect, which indicated positively that photons show *particle nature*. However, lest we get "carried away" by this apparent duality, I reiterate that the waves seen on the screen reveal ordinary geometric and physical behavior of waves meeting in space, coupled with the ordinary statistical wave-like distribution of many particles.

If we cover slit S_2 and examine the pattern on screen near P_1, the prominent parts of the interference pattern vanish, and one band seems to remain. This finding has attracted much attention, since it suggests that the "wave nature" of photons has evaporated. However, if we look hard enough we find *a faint interference pattern around this location*; we can see one more prominent band and several dimmer bands that become much fainter further away on the screen. This pattern is called *diffraction*, in this case single-slit diffraction, and *it alone is sufficient evidence for waves*. Nevertheless, diffraction from one slit is not as instructive as full two-slit interference, so that generally diffraction is ignored in quantum-mechanical experiments, and historically two-wave interference has enjoyed most of the attention.

Dispersion of light by diffraction occurs when a beam of jiggling photons encounters the edge of any opaque object, not just the edge of a slit. (Hence the edge of a shadow is never a sharp line.) This effect is also evident when light is passed through a very small hole, a fine grating, or a crystal. In fact, diffraction of light is visible when a coherent beam passes through a small hole; projected onto a screen, a circular interference pattern may be noted. If the beams are not coherent, color dispersion is also seen, but early two-slit experiments were too crude to reveal such subtleties. Nevertheless, in theory diffraction is an integral part of any two-slit experiment—since diffraction, even if subtle, occurs at each slit—so that the pattern on the screen should be called a diffraction/interference pattern. Ergo, the imputed "wave nature" of light has not all vanished when one slit in a two-slit experiment is covered! Moreover, the wavy (fuzzy) nature of dispersion—even when noted in two-slit experiments—can be predicted on the basis of jiggling without the imputation of wave-particle duality.

As I mentioned, modern two-slit experiments routinely include additional observations. One very important step in such experiments is to set up "particle detectors" which allow us to *know how many photons* traverse one or both slits. In strict terms, the screen itself is also a "particle detector," but in this context the particle detectors are very sensitive counting devices placed upstream near the slits in the barrier. Think of a person counting how many people entered through a door, or better yet think of two persons counting, one at each of two doors, with the refinement that we can tell which visitor passed through which door.

The startling finding is that *with a particle detector working at one or both slits, the prominent interference pattern vanishes*. In fact if we introduce any method by which we can tell which slit the photons pass through on its way to the screen, this item of evidence of wave-behavior disappears. It is even unnecessary for the experimenter (or anyone) to know what the particle detector has detected; as long as the device is active, no typical interference pattern appears. It seems that if the particle detector merely "knows" the path of photons, this "knowledge" changes their behavior. Given the inference that interference patterns signify waves, *it appears that if we examine light as particles by counting these particles, it does not act like waves.*

Moreover, if the particle detector is disabled or removed—if we stop detecting light as particles passing through a slit—*the interference pattern reappears*. Indeed if we turn off all devices that could count or tell is which slit the photon traversed, this manifestation of wave-behavior returns. It seems that a photon "knows" when we are looking for particles and when we treat it like a wave. If we seek particles, e.g. we count them, the photons seem to "know" to appear as particles and to hide their wave-behavior. But when we look for waves, the photons appear as a wave. How can a photon "know" when to show which of its two modes of behavior?

Even more baffling is the observation that when only *one particle at a time* is allowed to pass through the two slits, it shows the same agility, only more slowly. If we set the experiment to detect single photons and determine which slit a photon traversed, we find no interference no matter how many photons arrive. However, if we set the experiment to detect waves (by removing particle detectors), then each single photon, gradually and one at a time, contributes to the pattern until an interference pattern emerges. Does this mean that a photon *passes through both slits at the same time*? Does this also mean that *one* photon exits from the two slits in the form of *two waves that interfere with each other*? Does this also mean that one photon "knows" where previous photons landed (one at a time) so that it can contribute to an interference pattern? Does one photon "remember" where the previous one ended up?

Incidentally, one possibility is that photons interfere with each other like any particles when they bump against each other or repel each other. However we can generate photons so slowly that only one passes through the apparatus at a time, and yet the effects are the same as with many particles together. The difference is that more time is required to build an interference pattern.

These experimental results are easiest to elicit when the two slits are close together. However the interference pattern can be made to appear, vanish, and reappear as above *even when the slits are far apart*. It seems that one photon can pass through two widely spaced slits! It apparently "knows" immediately from afar when we are testing for wave-behavior, and again it seems to "know" where each previous photon landed so as to build an interference pattern. In other words, apparently individual photons can follow two widely separated paths at the same time to form waves that can interfere with each other, but only if we are unable distinguish which path was taken (by not detecting the photons before they land). We can therefore summarize such elaborate two-slit experiments as follows. *They imply that if we gain knowledge about how the particles got to the screen, we lose knowledge about their wavy behavior.*

However, do these observations really constitute a mystery within quantum mechanics? What if *any device able to detect individual particles disturbs the probabilistic distribution that is responsible for interference?* Think again of someone or something counting people who enter through a door, but also imagine that the entrants march in a very precise manner, e.g., in step with each other and at the same speed; in a word, they act coherently. In a well-designed experiment, this orderly behavior would have allowed interference patterns to emerge, but now in the process of counting, *the "counter" alters the behavior of the entrants.* The alteration—the *disturbance*, which is what it really is—can be subtle, but it is unavoidable and *it is sufficient to disrupt interference* once the entrants got inside. This disturbance has a name in quantum theory: decoherence, which in general is the result of contact with the environment. In this case the side effect of counting is "environmental decoherence," which entails *erasing interference.*

But why is this effect unavoidable? Please think in terms of small particles again: Even if they are in phase (monochromatic or coherent), these photons jiggle, and this alone blurs the lines on the screen. However, any device inserted in their path must somehow "see" the individual passing photons, and this device must itself be made of—and must include—jiggling particles!

This notion recalls the uncertainty principle and the concept that any such device must interfere with the behavior of what it is working on; here the "counter" must interact with particles being counted. At the same time, interference in waves of probability is a delicate process, and interference is easily destroyed (adding to the technical difficulty in two-slit experiments). Please recall that if we try to count photons passing by, we must first see them. To see them we might illuminate them sufficiently with light, but our light source emits jiggling photons of its own. If the photons we wish to count come in delicate waves, these waves "try" to interfere with each other, but we have just interfered with their interference! And if we use a very weak light (hoping to avoid disturbing the process), then we can't see the photon passing by. Worse yet, jiggling also enables photons to "slip through" unaffected when they were not counted.

A way of looking at this process is that the randomness jiggling is very fragile and easily disturbed by any observation. We can express this idea as a chain of events: The reason interference vanishes is that one or both waves of probability are disturbed. Waves of probability are disturbed because probability is disturbed. Probability is disturbed because randomness is disturbed. Randomness is disturbed because jiggling is disturbed. Jiggling is disturbed by any process of measurement in the environment. But why should a measurement at a slit be so disturbing? Because a measurement, which implicates detection of illuminated particles, requires collisions (interactions) with illuminating particles. Why then can we not simply calculate the effects of the collisions? Because they too are subject to uncertainties.

The effect of these disturbances can be appreciated with the diagram on page 22. The waves in the unequally long rays E and F account for interference, as they represent two conflicting probabilities for a photon to arrive at point P_2. If one of these waves is sufficiently altered by a measurement (or any observation) at one of the slits, the interference pattern is blurred out. I posit that this result is hardly astonishing, despite its reputation as an enigma! Incidentally, it should now be clear why I consider the uncertainty principle to be more basic than wave-particle duality. The former plays a key role in explaining the latter. As the "mystery" of the uncertainty principle is resolved by considering jiggling, a resolution of the "mystery" of wave-particle duality follows readily.

It is important to stress that even if a particle detector causes the interference pattern to vanish in a two-slit experiment, waves of probability still exist. However, these waves are no longer able to form complicated visible (and historically dramatic) interference patterns. Nonetheless, *diffraction* still may be discernible. In other words, the waves we have traditionally attributed to wave-particle duality *show interference only when they are sufficiently **undisturbed***. Waves of probability—even when two of them interact to elicit interference—merely obey the laws and geometry of the physics of waves. Interference is only one kind of wave behavior that is evident when certain conditions permit, including no significant disturbance by particle detectors.

One pesky issue often raised in discussions of wave-particle duality is this: Calculations of probabilities based on classical physics—essentially Newton's and Maxwell's—fail to anticipate the typical results of two-slit experiments, particularly when particle counters are in play. (Feynman for instance stressed this point; page 1-10 op. cit.) My response is to deduce and reason with the fragility of jiggling in mind. After all, Newton did not consider jiggling. In particular, inserting particle counters at the slits of two-slit experiments alters the subtle and unappreciated probabilities, and *that* is why classical equations do not account for this effect.

Here I wish to underscore another fine point. The notion that photons appear more densely spaced in some locations than in others can also be understood in terms of time rather than space. Let us focus on one region on the screen of a two-slit experiment where a new photon appears every few seconds. We then focus on another region where a new photon appears every few minutes. Thus we may find that 1000 photons reach the first region every hour but only 100 photons reach the second region every hour. Clearly the first region will quickly show greater photon density, but we determined this in terms of time.

Nevertheless, the first region will turn out to be on or near a crest and the second region on or near a trough of a wave. Then there is nothing astonishing if an interference pattern is generated by photons that are emitted gradually, even one at a time. Many photons in one region means high probability, while few photons in another region means low probability; we are witnessing a wave-like fringed pattern of probability, and the pattern is just an uneven aggregate of locations of particles generated over a period of time. Again, quantum mechanics does not require particles to bob up and down on waves or to turn into waves, but *when we make a record of when and where particles are likely to be found, we may find waves, which may or may not show interference*. As odd as it sounds, the probability is what bobs up and down; only probability makes the waves we see; particles remain particles, but they can be slowly amassed in patterns that appear wavy.

I dislike the term "frequency distribution," as the word "frequency" is used in different settings. However, if we agree that frequency means how often particles arrive at some given region of a screen, and if we see interference as alternating bands of high and low frequency, then we can say that the interference pattern is a kind of frequency distribution. This is just another method of describing and thinking about waves made of particles. It is not restricted to quantum mechanics, but we see the term "frequency distribution" applied to the results of two-slit experiments. The term "probability distribution" is better, and if that distribution is a series on bands, we can say that it is a wave—a wave of probability!

As noted earlier, light is not the only entity that shows wave-like behavior in two-slit experiments. Electrons behave similarly, including with regard to the disruption of interference in the presence of electron detectors. (Electrons require very narrow slits to show interference. The spaces between layers of atoms of crystalline nickel are just the right size, but the experiments are still more complicated than with photons.) Other beams, such as those made of protons, whole atoms, and even molecules, also show these effects. Interference is also found when more than two waves interact, but this complicates the situation needlessly and, in this context, it does not alter the interpretation of such an event. We still may see wavy patterns stemming from interference among waves of probability.

Incidentally, I mentioned (page 11) the "measurement problem" in quantum mechanics and the controversy over what constitutes an observation. For example, does setting up counters or particle detectors *without allowing humans to see the results* qualify as an observation or a measurement? Apparently yes, but I have a simple answer—a "functional" explanation: A process disruptive enough to erase wave interference suffices to be an observation.

Nonetheless the concept of waves of probability implies that in an observation or measurement, a probability changes instantly. For instance if an electron in an atom is found to shift or "fall" from one orbital to a lower-energy orbital—keeping in mind that each orbital represents a different probability of finding an electron in some location in the atom—the electron surrenders some energy, emitted as a photon, i.e. as a flash of light. Does this event possess a process that occurs in zero time? As implausible as it seems, does a probability change truly instantly?

It turns out that such a process can be examined experimentally, wherein an observation of an electron's shift is filmed over a millionth of a second (F. Pokorny et al, Tracking the Dynamics of an Ideal Quantum Measurement, Phys. Rev Lett. 124, 080401, Feb. 2020). The process during which an electron shifts its location and energy is a time-consuming evolution, during which the change of probability is actually wave-like and uneven. This finding makes sense and meshes with the idea of random jiggling.

Next I will demonstrate an advantage of thinking in terms of waves of probability, reusing the following diagram. Let me revisit the supposedly astounding possibility that one photon in a two-slit experiment can pass through both slits and appear in more than one location on the screen.

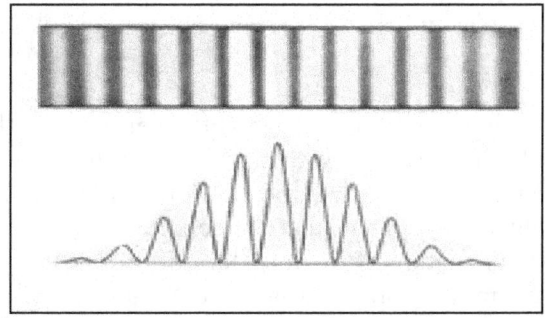

Here, as suggested by the diagram, a photon is more likely to end up in one of at least *ten* places on the screen, corresponding to about 10 crests in the wave pattern. Of course this pattern only emerges after many photons have landed, but it is formed simply because *each photon appears with sufficiently high probability in several sites.* It is not necessary to envision one single photon arriving in ten (or even just two) locations at once!

Given the nature of jiggling, it is even possible that some photons go around the barrier in an experiment. Worse yet, as I will discuss with tunneling (page 57) it is possible for photons to pass through the barrier anywhere, not just at the slits. These possibilities taint the mathematical analyses of two-slit experiments, but it still is not necessary for one photon follow more than one path to reach one point on the screen. (On the other hand, as I will also discuss, a modern mathematical approach to quantum mechanics [pioneered by Feynman] allows particles to arrive at a location by many different alternative paths, each with its own probability. The actual observed path is usually the one that is most likely.)

This scenarios address the claim that according to quantum mechanics, "one particle can land in more than once place at one time." As I suggest in the previous paragraphs, the reply is that before the particle is detected (e.g., before one dot appears on a screen) *its probability of arrival can be high in more than once place at a time.* In other words, a single particle does not have multiple simultaneous locations, but high points in its wave of probability certainly do. Thus no particle will land in more than one spot on a screen, but its wave of probability can have more than one crest on a screen, and that constitutes no mystery. Moreover, when photons are emitted one at a time, two separate detectors never detect two photons at once, except when it is possible for some emitters to generate two photons together.

Similarly, the observation that few or no photons appear in some places on the screen of a two-slit experiment may suggest that photons cancel each other out. The reply also invokes nothing peculiar: a few or no particles appear in places of low probability, including where waves of probability interfere. Waves can—and, in Feynman's approach, they must—cancel each other, but particles no not. (This is different from matter/anti-matter annihilation, to be mentioned later.)

There is yet another way of thinking about this issue, invoking the concept of quantization and recalling that photons are quanta: Since the source in a typical two-slit experiment emits photons which are the smallest possible units of light, one photon can never split in two in order to pass through the two slits. If instead the source emitted a wave that spreads out and passes through both slits at once, we should find more than one mark on the screen. Since we find only one mark per photon, we conclude that single particles pass through one of the slits.

Therefore, even when one photon is emitted and recorded, only one element of a wave can pass through two slits at once. That wave is a wave of probability, and the photon remains quantized and intact. Indeed if the behavior of photons (or anything used in the experiment) were not random, no wave-behavior would appear. I.e., if each photon were ejected from the light source the same way and nothing jiggled— if the initial conditions and subsequent paths were identical for each photon—then each photon would always land on the screen at the same location. If such were the case, a flashlight could burn a hole through a wall! Fortunately particles have an intrinsically unpredictable scatter; we can calculate and depict the probability for photons reaching some location, but we cannot be certain about the exact final location for one photon.

By the way, if a detection device is inserted into a two-slit experiment so that the interference pattern is eliminated, the addition of another device "downstream" restores the pattern. It seems that the jiggling can be allowed to resume, even after it was disturbed. Thus it is possible to observe "re-coherence," which sounds mysterious but which only signals that the conditions sufficient for wave interference have been restored. [10] However, given the nature of jiggling, the restored pattern will never be identical to the original. Worse yet, photons may simply "wander off" unpredictably. (We must not confuse environmental decoherence with "collapse of a wavefunction" upon a measurement or observation, which in the Copenhagen interpretation cannot be "un-collapsed." Collapse must be irreversible; different wavefunctions can collapse with the same outcome, and we cannot retrieve the past because 100% certain information on position and momentum never existed in the jiggling realm of particles.)

A possibility exploited in science fiction is "beaming" people or objects across space. Does quantum mechanics allow such "teleportation?" No, but "quantum teleportation" limited to teleporting quantum states is actually feasible, and indeed it is being applied is real situations. This fact raises the question whether quantum teleportation represents another mystery, which is important enough that I devote a detailed chapter to it (page 45 on). However, that discussion requires an exploration of entanglement first, which appears in the very next chapter.

---

[10] P. Kwiat, A.M. Steinberg, and R.Y. Chiao, "Observation of a "Quantum Eraser": A Revival of Coherence in a Two-photon Interference Experiment," Phys. Rev. A45, (1992): 7729-7739.)

Anyway, despite the striking results of two-slit experiments, I wish to point out that these represent rather extraordinary and contrived situations. Besides the need for a source of suitable beams, a strict set of conditions must exist for interference patterns to be manifest. Both slits must be open, the particles must be small, and no particle detectors can be active at the slits. The argument can therefore be made that wave-particle duality is largely a laboratory curiosity with little real significance. In other words, since the interference found in wave-particle experiments is so unnatural and artificial, it cannot be meaningful. Such misgivings about two-slit results miss the point. I re-emphasize that the "wave" part of wave-particle duality is overstated, no matter how it is revealed. The duality exists only between two aspects behavior of the same particles. Under some circumstances, as after an observation is completed (a spot appeared on a screen) or a particle counter is active (no interference pattern), wavy probabilistic behavior is not evident; we call this "particle behavior." Under other circumstances, as when repeated observations at various locations are completed, the probabilities for finding particles at these locations form wave-like patterns; we call this "wave behavior." Is this really a mysterious duality, or do we merely see particles whose final locations can sometimes be recorded so as to reveal wavy patterns? Practical uses of wave-particle duality exist, but they are merely applications of wave equations—such as those of Schrödinger, as I will mention below—to describe one aspect of the behavior of particles.

Let me be clear (even if repetitious): these waves *do not describe or replace the intrinsic properties of particles*. They describe and reveal *the likelihood of finding particles* at various locations. Let me also touch on two related but easily misused terms: "Wave packets" are bursts of waves of probability, concentrations in the chance of finding particles or sites of high "probability density." "Matter waves" are also known as de Broglie waves in his original hypothesis (page 16) and were used to explain diffraction. I enfold them into the concept of waves of probability, with no extraordinary properties.

Please recall the reputedly enigmatic feature of wave-particle duality that we find *either* wave-behavior *or* particle-behavior but never both at one time. I posit that the explanation is simple: at any one time we *can* either disturb particles enough to erase interference or we *cannot* do so. And as for the mystery of any one particle traversing one slit *and* the other slit, of course there exists a probability for one *and* for the other to occur (and even for neither to occur). Indeed a recent very complicated experiment [11] shows images of the wave-nature (interference pattern) and particle-nature (quantized photons) of light at the same time. Even if this is substantiated, the interfering waves are waves of probability of location recorded at "ultrafast" moments in time.

One inspiration for Schrödinger's wave equations was to explain the results of two-slit experiments, particularly interference. The classical counterpart of these equations is Newton's $F = ma$, itself a brilliant advance in physics but one that could not account for the observations on quantum interference. Of course Schrödinger's equations are vastly more complicated, and yet the simple arithmetic feature that allows for wave interference is that the amplitudes in these equations can be negative; basically, interference can occur "on paper" when negative and positive amplitudes cancel each other. Thus by solving Schrödinger's wave equations, we retrieve what we see in quantum experiments: events suggesting wave-particle duality. With these equations we also anticipate another notable finding: Experiments and quantum calculations on particles only reveal likely—rather than exact—outcomes, reflecting a basic uncertainty; the best they can do is provide probabilities. (Please see page 13 on determinism.)

In my view only three ideas are needed to dispel the alarming reputation of this wave-particle duality. One, there is delicate jiggling. Two, when any two waves meet in space, interference patterns may emerge, but these patterns are easily disturbed. Three, after passing through a slit, the points at which jiggling particles arrive form fuzzy and wavy clusters on a screen. Nonetheless, I have devised the

---

[11] L. Piazza et al. *Simultaneous observation of the quantization and the interference pattern of a plasmonic near-field.* Nature Communications, Article number 6407, March 2, 2015. (Note their figures 3 and 4.)

following tabular summary on the standard traditional interpretation of wave-particle experiments, on the concept of waves of probability, and on my "jiggling" interpretation:

Two-slit experimental steps	Observed results	Standard interpretation	Jiggling interpretation
Both slits open and no detector(s)	Interference pattern clearly seen	Wave-particle duality. Wave behavior and no particle behavior.	Wave pattern created from particle behavior. The wave is a wave of probability. No duality.
One slit closed	One fuzzy line	Wave-particle duality. Particle behavior and no wave behavior.	Diffraction, a wave pattern. A wave of probability. No duality.
Both slits open but particles released slowly one at a time	Interference pattern emerges gradually	Wave behavior but one photon interferes with itself.	Wave pattern created from particles one at a time. The wave is a wave of probability.
Both slits open but detector at one of the slits	Two fuzzy lines	Particle behavior and no wave behavior.	Diffraction, which is a wave pattern. Waves of probability. Counters disturbed the interference, ergo no interference pattern.
Both slits open but large objects used	Two lines	Particle behavior and no wave behavior.	Wave behavior exists but obscured.
Conclusion		Wave-particle duality. Wave and particle behaviors exist but only one is visible at one time.	Particle behavior but locations distributed by probability. That distribution can be wave-like, including in an interference pattern.

To make the conclusion (in the above table) more current, we note that in October 2012 the Nobel Prize in physics was awarded to David Wineland and Serge Haroche, both investigating quantum mechanics. The work of the latter is particularly germane to wave-particle duality. Working with other physicists, Haroche was able to trap and count individual photons while delaying their decoherence.[12] At the same time he noted wave-like activity in a large atom that had interacted with these photons. This activity still represents waves of probability, and yet even under these conditions the photons clearly retained their particle-like nature. More recently, in the words of scientists at the Center for Quantum Technologies in Singapore (P. Coles et al., credit Timothy Yeo/CQT, National University of Singapore), "wave-particle duality" is simply the quantum "uncertainty principle" in disguise, reducing two mysteries to one.

On this note I turn to the next main section, dealing with quantum entanglement and with the mysteries ascribed to this phenomenon.

---

[12] See Bibliography starting on page 71, section with citations related to Haroche's Nobel Prize. In particular see figures 4 and 6 in Haroche, Brune and Raimond: We see waves, but they are *waves of probability*.

Quantum Entanglement and Several Postulates

HARMONY, PROBABILITY, CURLED-UP DIMENSIONS and SUPERDETERMINISM

As the title and subtitle of this section indicate, I invoke several postulates to deal with the mystery associated with quantum entanglement. Before presenting these I must discuss quantum entanglement—called just entanglement for brevity—in some detail. I have additional reasons to give this topic close attention. For one, this field is turning out to be quite practical, as entanglement may find significant applications in modern technology. It is also a very thought-provoking discovery, as entanglement may be the most "non-classical" and counter-intuitive aspect of modern science. For another, the nature of entanglement has instigated what I consider to be very important—and even the most intellectually fascinating—disagreements, debates, controversies and difficult experiments in the history of science. These issues were brilliantly championed by "giants" such as Albert Einstein and Niels Bohr, and as of today the issues still have not been definitively settled. It is therefore also no surprise that a simple explanatory picture for entanglement—one solitary postulate—is difficult to find.

I shall begin with another analogy. Niels Bohr is a clever gambler, and Albert Einstein owns a casino. Albert suspects that Niels is cheating at dice, since Niels wins too often. Niels explains that the dice are "entangled" in such a way that whenever a 3 comes up in one die,[13] a 4 is likely to come up on the other, because when the first die shows 3 it can "tell" the other to show a 4. In other words Niels avers that the dice communicate with each other, thereby explaining the correlation that characterizes their "entanglement." Thus he often throws winning numbers.

Niels also boldly claims that this communication is immediate, even at a great distance between the dice. For instance, he predicts that he can quickly win more often than expected even if each of the two dice is tossed on separate tables, or in separate casinos, or on separate continents, or even on separate planets! No matter how far apart are the dice, a 3 in one die often correlates with a 4 in the other without delay.

Niels also points out that each die alone shows nothing noteworthy or sinister. The correlation only appears in the conduct of the pair. Moreover, Niels stresses, one throw of the pair of dice is not significant. Since each result is still subject to the laws of statistical probability, many trials are needed to draw conclusions. Of course Niels knows enough not to win too often, and he admits that if a 3 or 4 comes up he allows the dice to communicate only sometimes. Albert knows that even a slight edge is significant, but he is not convinced by Niels' explanation. Indeed to Albert all this sounds preposterous and "spooky." He avers that since dice cannot possibly communicate instantly, they must somehow be "pre-programed"[14] to act that way. It seems to him that Niels' explanation is therefore "incomplete," and in particular Albert believes that the dice are secretly altered before every game so that the result on one die complements the result on the other. Moreover, Albert infers that this pre-programed condition of the dice—a "hidden" state—does not change while they are in flight. In his view, this is why the dice need not be close together for their correlation to be instantly manifest. They need not "tell" each other anything instantly even if they could, which to him is absurd anyway.

Still, Albert realizes that Niels's boast and apparent success have a far-reaching implication. If Niels is right, it means that two objects—at least when entangled—can indeed communicate with each other instantly across vast distances! But how can that be? How can a meaningful message travel across space without delay? The transmission of all other factual messages is limited to the speed of light, so how can entanglement be exceptional?

---

[13] The singular of dice is die.

[14] Also spelled pre-programmed.

Albert further reasons that if objects can be pre-programed, then they must hold real and factual information within themselves before their entangled behavior is actually observed. He is determined to resolve the mystery: Is Niels using dice with hidden secret information encoded in them, or are dice really capable of "spooky" long-distant incredibly rapid communication?

Albert turns to his friend John Bell for help. John carefully considers the problem and deduces that if the entangled dice are initially pre-programed and locked into a mode of cooperative behavior, they should behave somewhat differently than if can they "talk" to each other. In the latter case, their correlation should be statistically tighter than in the former case.

John concludes that the issue boils down to detecting an excess statistical correlation, one which would implicate communication rather than pre-programing as the explanation for the mysterious features of entanglement. John even suggests how an experiment could be designed to detect such a correlation, though he knows that this will not be easy. Privately John hopes that such an experiment will uphold Albert's views, if only out of respect for Albert's reputation and uncanny ability to reason accurately.

But Albert is worried. From what he has seen, he cannot deny that the dice behave in an entangled manner—after all, Niels has amassed a fortune in winnings—and yet Albert cannot accept the possibility that widely separated things can communicate instantly. But what if the dice fail to behave as if they are pre-programed? What if John's excess correlation is actually found in experiments? Then Albert may have to concede that the dice do show "spooky action at a distance," as incredible as it sounds and as rational as Albert's skepticism seems to be.

Finally John's experiment is carried out, and astonishingly *it appears that the members of a pair of entangled dice can show excess correlated statistical behavior.* In this context "correlated" means involving both particles together; a synonym is "inter-dependent." "Behavior" is something measured or observed in a real experiment. The point is that Niels may be right!

This ends the analogy, and there really does exist a set of crucial experimental observations, ones which have been duplicated in many sophisticated physics laboratories, but ones which were unknown to, and unanticipated by, classical physics: Photons which are created in one process can be entangled, *and the members of an entangled pair of such photons may show strongly correlated physical behavior that can be detected instantly no matter how far apart the two particles are located.*

This inter-dependence does not exist without entanglement, and it exceeds what chance alone would allow. (Of course some correlation is expected purely by luck, and well-designed experiments take that possibility into consideration in their statistical methods.) Particles showing such behavior—statistically significant excess correlation—are said to be in a distant entangled quantum state. In reality, and not just in an analogy, Niels Bohr and not Albert Einstein may be correct! Gasp!!

Many scientists have worked on this issue for decades, and the effort continues. Schrödinger first used the term "entanglement" in his native German as "Verschrankung." [15] He considered entanglement to be the primal feature of quantum mechanics, though others (e.g., Heisenberg and Feynman) stressed the pivotal roles of other aspects of quantum mechanics, such as uncertainty and wave-particle duality.

I will come back to how particles might attain their entangled "state," but let me turn to some essential technicalities I have omitted so far. A behavioral property of photons that has revealed entanglement is their intrinsic angular momentum called "spin" that can be labeled as "up spin" or "down spin." Actually the spin of particles is relatively difficult to work with; photons have another more "user-friendly"

---

[15] Though experiments on entanglement are difficult to perform, entanglement itself is not a rare state in nature. For example the hardness of a rock depends on entanglement.

property, polarization,[16] but for the present purpose spin is more illustrative. Also, spin was considered earlier than polarization to study entanglement, and electrons also exhibit spin.

In the absence of entanglement, the spin of one photon—be it "up" or "down"—is unrelated to the spin of another distant photon. However, let us imagine that *two entangled photons* are sent in opposite directions until they are very far from each other, and their spin is not disturbed during their journeys. We find that the spins of these photons are promptly correlated. For example, an experimenter who accompanied one of the photons determines that at noon on January 1, 2012, this photon had "up" spin. Another experimenter, who accompanied the other photon and agreed to make his determination one nanosecond after noon on January 1, 2012, finds that this other photon had "down" spin. (We can ignore problems of clock synchronization.)

The experimenters had also agreed to repeat the experiment on many pairs of similarly entangled photons and to keep accurate records for subsequent comparison. On such a comparison they found that often—significantly more often than by sheer chance—whenever one experimenter found "up" spin, the other experimenter found that the other member of the entangled pair had "down" spin. Similarly, often when one experimenter found "down" spin, the other experimenter found the opposite. Moreover, this correlation appeared within a nanosecond no matter how great the distance between the sites of the determinations.

These observations are disconcerting, because they insinuate that the two entangled but widely separated photons "communicate" with each other, and they do so incredibly fast. Thus somehow when one photon reveals its spin, it immediately seems to "talk" to its distant entangled mate and instruct it to reveal the opposite spin. Ergo, one measurement here influences another measurement far away by means of a mysterious super-messaging.

However, *there is an alternative implication*: Perhaps the photons did not "talk" to each other, but they "knew" prior to the measurements which way to spin. That is to say, the response of each photon to a determination of spin can be explained by some kind of pre-programing that was achieved during the process of entanglement. Such pre-programing provides innate synchronized information which each photon carries to the site of the determination, and the observed correlations require no further communication between the photons. This information does not alter the individual behavior of each photon when studied singly, but it affects the pair. Moreover, the information may not be readily evident; it may be hidden till it is revealed.

The explanations proposed by scientists for entanglement fall into two categories. One category is based on the principles of quantum mechanics upheld by Niels Bohr (and others) and compiled into a system called the Copenhagen interpretation.[17] This approach avers that the instantaneous correlations between entangled widely separated particles do reveal a form of communication between the particles, and no pre-programed information exists. The opposing view, sustained by Albert Einstein (and others), insists that no physical correlation can span a distance instantly, but that particles can be prepared in advance to show correlated behavior later. That preparation consists of some pre-programing with information, and this information may be hidden. No matter what correlation is found, communication is not involved.

A key element of this scientifically profound issue is Einstein's position, based in his theory of special relativity, that all physical processes progress at a definite maximum speed, namely the speed of light,

---

16 Polarization is a property of light after it passed through a certain crystal or lens, where it is called linear polarization. In our context polarization depicts the spatial orientation of the electric field of a photon.

17 The Copenhagen interpretation was developed mainly by Bohr and Heisenberg in the 1920's and 1930's. Bohr's laboratory was in Copenhagen. This interpretation is also known as the Bohr-Heisenberg interpretation. See Hughes, pp. 214, 306-311, in Bibliography starting on page 71.

which means that separated but causally linked events require finite and significant amounts of time. Ergo, meaningful communication across space, including the transmission of information, always takes a while and can never be instantaneous. Thus the length of the space interval between the measurements of spin described above is critical because if that gap is large, a nanosecond may not be enough time for a message to get from one photon to the other, even if that message is sent at the speed of light.

Relativity has a term for a distance that is so great that it cannot be crossed at the speed of light in a given time period. Such a distance is "space-like," in contrast to "time-like" distances that can be covered in the allotted time at the speed of light. When a process is reputed to be space-like, Einstein vehemently averred that it is a case of "spooky action at a distance." Since such "action" exceeds the speed of light, it is said to be "superluminal," and in this case there is superluminal communication between photons regarding their spins, in defiance of special relativity. Einstein hence insisted that instant communication across a space-like separation is physically impossible except by afore-described pre-programing. (Transmissions that do not exceed the speed of light are "subluminal" and do not violate special relativity.)

Before going further I must introduce two other important terms and the concepts to which they allude: locality and non-locality. The corresponding adjectives are local and non-local. In brief, local physical events—those that are near each other or act as if they are adjacent—can easily influence each other, but *according to special relativity such "local" events cannot instantly influence events that are far away.* The concept of locality embodies the relativistic principle that no physical influence, measurable effect, or meaningful information can arrive from afar faster than at the speed of light. The implication is that only local influences—local causes—can be responsible for a physical event. By virtue of special relativity, causality cannot be superluminal. It cannot be non-local.

The pertinent point is this: *According to Copenhagen quantum mechanics, the two entangled particles are actually non-local, permitting superluminal influences between events.* Here the term "non-local" is practically equivalent to "space-like-distant," and a non-local wavefunction[18] presumably embraces both particles concurrently despite their great separation. The compact wording of this issue is that the Copenhagen interpretation maintains that *"quantum mechanics is non-local."* A simple model for non-local behavior by entangled particles is acting as if they are adjacent, or as if there were no space between them even, though they may actually be very far apart.[19] I will return to this point in presenting my postulates to clarify entanglement.

Of course Einstein denounced the non-locality espoused in quantum mechanics. We say that he and his supporters subscribed to a "local" paradigm to account for immediate distant correlations between entangled entities. A reverse wording may be clearer: If quantum mechanics in general and entangled particles in particular were local, then instantly correlated behavior between these particles—spooky action at a distance—should never be observed. Alas such behavior may be observed experimentally.

Einstein's original German term for "spooky action at a distance" was "spukhafte Fernwirkungen," which literally means "spook-like distance-effects." Therefore the traditional English equivalent is somewhat inaccurate, but the idea is that something significant which supposedly can cover a distance impossibly quickly is not science but magic. I.e., it is a deep mystery. Einstein voiced this opinion in many ways, but a renowned forum was in a crucial 1935 physics article co-authored with Podolsky and Rosen called

---

[18] A wavefunction is an equation used to fully describe something (e.g. a particle) for the needs of quantum mechanics. Wavefunctions are key elements in the math and theory of quantum mechanics, particularly in the approach developed by Schrödinger and Bohr. I discuss wavefunctions on pages 18 and 28.
[19] See McFadden, p. 200. "Particles should demonstrate correlations...instantaneously communicated to all parts of the quantum system, as if space did not exist between them." See Bibliography starting page 71.

the EPR paper or "EPR" for short.[20] However, their argument was more subtle than just a restatement about space-like processes. EPR averred (though not in those words) that quantum theory needs additional ingredients to explain what looks like "spooky action at a distance."

EPR was not the report of a completed experiment; by 1935 no techniques for laboratory verification had been developed. EPR was a "what if" dissertation, a "thought experiment" ("Gedankenexperiment" in German) for which Einstein was famous. EPR can be seen from two points of view. One is "historical," based on the arguments voiced in 1935. The logic was circuitous, but it was then "state of the art." The other vantage is the "modern" one, in which the issues raised in EPR are tackled using insights and methods that are currently available. An important point is that these modern developments have allowed at least a partial or tentative resolution of the controversy in the behavior of entangled particles.

Historically, the thrust of EPR is an apparent deficiency in quantum mechanics, as evident in the title of the paper, "Can Quantum-Mechanical Description of Physical Reality Be Considered Complete?"[21] The primary issue in EPR was not the nature of entanglement. (The word "entanglement" is not in the EPR paper, and this phenomenon was merely a tacit part of the method.) Rather, EPR was an important early chapter of the debate between Einstein and Bohr: Is quantum mechanics a competent scientific theory? And do objects—represented by particles—possess physical quantities before these quantities are actually measured? In brief, Bohr argued that such properties do not exist prior to their measurement. Meanwhile, Einstein held that unobserved properties really do exist, even if these are not measured, and quantum mechanics fails to recognize them because quantum mechanics is incomplete. These omitted properties or quantities, possibly preprogramed, later became known as "hidden variables."

As scientific thinking evolved in the second half of the 20th century, the focus shifted dramatically: The main preoccupation today—and hence the mystery I aim to explain—*whether quantum mechanics in general, and entanglement in particular, are non-local.* In short, does superluminal communication between distant physical entities exist? EPR's questions about whether quantum mechanics is complete and whether objects possess real quantities faded into the background, in part because these issues seemed to be mainly theoretical and even philosophical, and in part because quantum mechanics as it stood appears to be a reliable physical as well as mathematical theory. That is to say, in modern thinking it matters little whether quantum mechanics has overlooked hidden variables, but it matters very much whether entangled particles are really capable of incredible superluminal non-local communication!

At this point I must say more about research on entanglement. Next is a diagram of a basic experiment for studying entanglement in the context of quantum mechanics. A "photon source" emits two beams of entangled and polarized photons; they are entangled thanks to the way the source was designed, but they are also polarized vertically (at zero degrees, as on a compass) when they are emitted from this source. (Polarization rather than spin is used more commonly in such experiments.) Here photon I, moving to the left, encounters a vertical filter while photon II, moving to the right, encounters a filter set at 30 degrees. Since we assume that photon I has passed through a zero-degree filter, we expect that its entangled mate, photon II, behaves *as if also has zero-degree polarization.* (These two photons can switch roles.) One detail that will be critical in my subsequent discussions is that in usual experiments the selection of filter settings must be completely random, lest a bias be introduced that can invalidate the outcome.

---

[20] Podolsky and Rosen were post-doctoral research associates at the Institute for Advanced Study at Princeton University. Rosen also is associated with the "Einstein-Rosen bridge," a theoretical space-time construct in general relativity.

[21] A. Einstein, B. Podolsky, and N. Rosen, "Can Quantum-Mechanical Description of Physical Reality Be Considered Complete?" Phys. Rev. 47 (1935): 777-780.

36

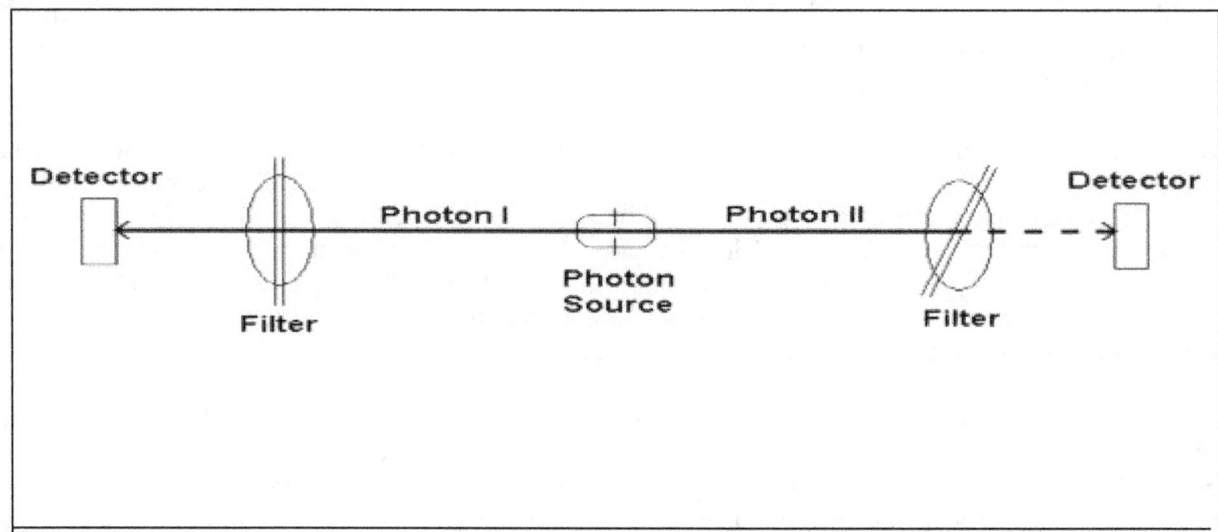

In this diagram we assume we create a pair of entangled polarized photons at a photon source. One photon is aimed to the left and the other to the right; these are labeled I and II respectively. Each photon encounters a filter. The diagram shows a vertical left filter and a 30-degree right filter. A detector senses whether or not a photon has passed through a filter. The interrupted path after the right filter indicates a reduced probability of transmission.

This diagram suggests a basic question about the photon source: What happens here when photons become entangled? Or in quantum-mechanical terminology, how did these photons acquire their entangled quantum state in the photon source? Several techniques are available to achieve entanglement: Parametric down-conversion (this method and others rely on the laws of conservation of energy and momentum), atomic cascade decays, and high-energy collisions between certain subatomic particles. To protect the validity of the experiment, the establishment of entanglement must also be free of bias with respect to the angular direction. (Entanglement may occur naturally, but this has not been harnessed for experimentation. Also, when there is entanglement, the polarizations are usually expected to be parallel [depending on the method], whereas spins are expected to be opposite under these conditions.)

In this context I aver that jiggling helps envision what happens, and jiggling can dispel the mystery of entanglement. I posit that *the process of entangling the behavior of two (or more) particles introduces a slight but significant degree of harmony into the jiggling of each member*, and this harmony manifests itself in the correlated microscopic behavior of the members of the entangled group. In the photon source the members of that pair are very close to each other as they attain their communal properties. Therefore there is no space-like separation between them, no superluminal process is involved in attaining their entangled state, and—by virtue of enough harmony in jiggling—*no subsequent communication between them is needed.* Think of two skilled singers close to each other as they sing one soft note. They can stay in harmony with each other (as much as possible) after they move apart and no longer hear each other.

Of course if the polarization of just one photon is determined, its direction is found to be random, as is its jiggling. Indeed when particles are studied individually, any potentially entangled behavior is not evident, simply because any harmony in their joint behavior is not revealed until they are compared; such behavior is only local or regional (and exists only as long as their entanglement endures). Think of the singers again: We cannot tell whether the duet is singing in harmony by listening to just one member; the other member's note could be anywhere on the scale. We will find out if they were in tune only by directly comparing their two vocal outputs.

A critical question comes into focus now, considering what experiments reveal: Why does entangled behavior of distantly separated particles appear to be non-local? I can picture an answer by invoking jiggling: As long as the particles remain entangled, *some harmony in their jiggling remains* even if it is altered somewhat by what happened to the particles in flight. Thus if one or more of the particles passed through filters which alter their angles of polarization, an altered harmony is induced, but *it is sufficient for them to continue to show correlated behavior without any intercommunication.*

The exact previous state of the harmony is irrelevant; my thesis is that their joint behavior has been reset by the filters, but the duo continues to behave as a harmonious unit. This unity can be reflected mathematically in the wavefunctions of entangled entities. In quantum theory entangled particles share a mathematical description, their common wavefunction. See details on page 18. (Wavefunctions can be written as superpositions, and in cases of entanglement, these superpositions are "non-separable." This even applies to entanglement among more than two particles.) In other words, I subscribe to Einstein's position that pre-programing has occurred in the photon source, and this pre-programing takes the form of a sufficient degree of harmony in the jiggling of entangled photons, and it remains evident when an experiment is completed. *I assert that harmony in the particles' jiggling can account for their distant but instant correlated behavior.* This concept easily shows how Einstein could have been correct; in my view the missing element in past quantum theory—the "hidden variable"—is *harmony in jiggling.*

Jiggling explains other aspects of experiments on polarization: It is why actual experiments do not give ideal unambiguous results; by analogy this is why the two singers are unlikely to be exactly in tune. Upon any measurement the random feature of jiggling becomes somewhat overt, so that the experimenters can neither control nor predict exactly how particles behave; not all randomness has been eliminated by the harmony. My position therefore is that if you can (1) accept jiggling as a constant but delicate vibrational random motion, and if you can (2) picture two or more entangled particles whose jiggling currently is in some state of harmony, and if you can (3) think of an entangled group as an adequately harmonized entity, then the salient features of entanglement are intuitively predictable. This brings me to one of the postulates I use in thinking about the mystery in entanglement, my HARMONY IN JIGGLING POSTULATE:

Entanglement arises from a delicate and subtle but sufficient state of harmony in the otherwise random jiggling of two (or more) particles.

Since the required pre-programing assured enough harmonious jiggling, members of a pair of entangled photons can show instantly correlated physical behavior in real experiments. Let us apply this concept to the diagram on page 36. We expect that any photon I with zero-degree (vertical) polarization will very likely pass through a zero-degree (vertical) filter, though of course the probability will not be quite 100% in part because photons jiggle. Now what happens when photon II, which also shows zero-degree polarization by virtue of entanglement, encounters its filter (filter II) which is not at zero degrees but instead set (randomly) at 30 degrees? The probability of traversal should be reduced, and thus we should find less likely transmission. That is to say, when the filters are misaligned, there is a probability much less that 100% that a photon with zero-degree polarization will get through a 30-degree filter. The diagram shows this by an interrupted path for photon II (between the oblique filter and the detector).

However, the key experimental finding is not just that misalignment of the two filters causes a reduction in the rate of detection of the photons that get through; this is easy to envision. Nor is it surprising that more misalignment of filters causes more reduction in the rate of detection. What ultimately commanded the attention of the world's scientific community is a seemingly small detail: *exactly how much reduction is found at any particular amount of misalignment.*

At this point we must know that a now-famous physicist, John S. Bell, in 1964 published a startling and profound insight: *The exact amount of reduction will not be the same if Einstein is right—polarized entangled photons behave according to how they are pre-programmed regarding—as it would be if Bohr is right—polarized entangled photons appear to communicate in a non-local manner.* According to Bell's derivations, this difference should manifest itself as a *statistically stronger correlation in their behavior in the latter case.* In terms of our singers, there is a probability that they remained in tune after separating, but if instead they relied on communication to attain their harmony, *that probability would be somewhat greater.*

This prediction, misnamed Bell's theorem (it's not really a theorem), has turned out to be a monumental development in quantum physics. It meant that the feud between Einstein and Bohr might be settled by a presumably straightforward experiment: Which scenario, pre-programed information or superluminal communication, is supported by the amount of observed reduction in how many entangled photons pass through misaligned filters? In short, is the observed correlation between entangled particles statistically excessive? I posit that harmony in jiggling may explain why some experiments do reveal such excessive correlations, implying that entangled entities can behave non-locally.

Admittedly, a problem with my theory is lack of direct experimental revelation of such harmony. Still my postulate can account for Bell's theorem via the assumption that the degree of harmony is sufficient and durable enough to explain the experimental results. One feature of such correlation is that it conforms to the cosine-squared function in Malus' Law of optics, and it is possible to devise a harmony-based model that obeys that law. Moreover, this assumption does not preclude communication between entangled particles, just as harmonized singing does not prevent the singers from messaging to each other to improve their vocal harmony. Therefore I now point out that even if such superluminal communication exists, *it does not necessarily include the exact note that the singers seek.* At best it can only be *an approximation.* Why only an approximation? Because their harmony must be at least somewhat imperfect. And why imperfect? Because it is affected by jiggling! In other words if *any* message passes between the singers, it is subject to at least some uncertainty and probability. Likewise, in experiments on entanglement *any communication between particles must be affected by jiggling.* I will return to this point, but I should first detour into some additional technicalities.

Despite the ongoing spate of such experiments after Bell's publication, settling the issue turns out to be much more elusive than Bell's theorem suggested. The problems are essentially twofold: It is difficult to work with jiggling particles moving close to the speed of light, and, more importantly, the results are indecisive because they are open to different interpretations as to whether they really prove that quantum mechanics is non-local. [22] At times the controversies over these results are even more divisive and acrimonious than those between Einstein and Bohr, even though the experiments are still becoming more sophisticated and precise. It seems that every refinement sooner or later is met by refined objections or criticisms.

Other properties besides polarization have been studied, and other particles besides entangled photons have been used, but with the same controversial results. A recitation of examples of current research and their details would fill volumes, and it is not essential here. (E.g. recent experiments with three or more

---

[22] The issues in these experiments open to challenge are called "loopholes." E.g., the detectors typically detect fewer photons than were launched by the source. Does this discrepancy discredit the results since only a sample of a population was studied? Also, given the enormous speed of light, are the members of an entangled pair far enough apart so that there is insufficient time for superluminal messaging? A 2015 experiment using photons and electrons, cited on page 72 of the Bibliography, claims to have closed all loopholes: Entangled electrons were far apart, the test angles were selected randomly, and most entangled pairs were recorded. However, the method of entangling the pairs of electrons (using entangled photons) appears to allow harmony, and *the results still entail statistics and probabilities.* The authors (on p. 7 of their report) admit that all alternative theories cannot be excluded, fundamentally because it is impossible to prove when and where quantum entities came into existence. I touch on this fact on page 60 regarding how quantum particles behave; they behave probabilistically.

entangled photons at a time yield additional information but are very complex, and I cite [below] a very sophisticated recent experiment with electrons.) In general however, the interpretations of the results are still open to dispute, even though the "loopholes" in these experiments—please see footnote—are being "closed" through better techniques and more advanced analyses. I will come back to the loophole issue where it fits into my context.

On the other hand, even if not every scientist is convinced and no experiment is decisive, *the outcomes of these experiments largely do support non-locality*. Therefore let us assume for now that indeed Einstein was wrong (gasp again!) and that nature does permit "spooky" superluminal behavior; extraordinary correlated physical behavior does exist. In other words perhaps entanglement itself can be non-local, and hence so can quantum mechanics in general. Therefore I now return to the point that my harmony postulate itself may not elucidate conclusively whether particles communicate. Here I have an additional postulate in mind, and it again involves jiggling. However it is more intricate, as it draws on features of special relativity.

For this purpose I focus on the randomness in jiggling that persists even in the face of any harmony. As it stands, all experimentation on entangled behavior is basically statistical, in the sense that focusing on one pair of photons does not tell us enough and that there is at least some uncertainty for any particular experimental result. Ergo, to glean a satisfactory and meaningful result, *we must consider the statistical behavior of many pairs of photons*. (The same applies to recent experiments with triplets of photons or even larger groups.)

It is important to stress that when photons pass through a polarizing filter set at some angle, most of these photons acquire that angle of polarization, but their behavior is still influenced by probability, and all randomness is not eliminated. This is because they are still jiggling, though this jiggling, delicate as it is, has been altered by the filter. Indeed the photons' orientations in space are described by waves of probability of the kind we discussed in the previous section; i.e. waves reveal where the photons are likely to be found if we bother to seek these waves.

Let us assume that these photons passed through a filter set at zero degrees. (I still envision a compass-like system of coordinates with "up" and "vertical" at 0 degrees; this is arbitrary but easier to apply.) We can then draw a wave of probability that favors a vertical orientation. In terms of location, these photons are most likely to be found in a plane at zero degrees. They are less likely to be in a plane that is, say, 5 degrees from the vertical, still less likely to be in a plane 10 degrees from the vertical, and very unlikely to be in a horizontal plane at 90 degrees. I re-emphasize that each case can still be described and depicted by a wave of probability.

If the photon which traversed a vertical filter next passes through another filter which is horizontal, its horizontal jiggling is mostly but *not* totally eliminated, so that now—again analogous to the case of location of the photon—the probability of finding a photon after both (perpendicular) filters is just nearer to zero. That is to say, the chance for finding a photon at the location of the detector ("downstream" from the two perpendicular filters) is low *but it is not nil*.

With respect to such probabilities, the photon detector in an experiment on entanglement is the same as the particle detector in a two-slit experiment. For that matter, it is the same as the particle detector in any quantum physics experiment. It merely provides data for determining the *probability* that particles are detected at this location. In general, as I maintain, this probability forms a wave-like pattern, and it ultimately stems from the randomness of jiggling. Furthermore, in experiments such as the one envisioned in EPR as well as such as those actually used to test Bell's theorem, the results are unknown until data from both members of entangled have been assembled. However *that step, the compilation of probabilistic measurements, does not require superluminal communication*. In fact it can be done in a quite mundane and leisurely manner long after the measurements have been done.

Please recall the above long footnote on loopholes in the experiments which demonstrate a violation of a Bell's inequality and which therefore imply the necessity for superluminal communication to account for entanglement. The cited work is an example of dealing with the technical problems I mentioned earlier that confound research on entanglement. The authors admit that despite running their experiments for many hours, only 245 instances successfully provided "high-fidelity" entanglement between distant electron spins. The obvious question is this: Is the "failed entanglement" in unobserved particles an integral feature of entanglement whose exclusion from the experiment biases the conclusion. Perhaps high fidelity is exceptional and low fidelity is the norm when particles show their correlated behavior.[23]

These points provide another escape from the dilemma that quantum mechanics apparently sanctions superluminal transmission of information (non-locality) while special relativity forbids it, particularly if direct observation of harmonious jiggling in entanglement may be impossible. Here I draw upon the concept that *no useful information is being conveyed via entanglement*. A distant observer does not know for certain (until notified) whether the angle at his filter is the same or different than the angle at the other randomly selected filter, and a compilation of many trials is needed. In effect, *since no certain or definite knowledge is transmitted superluminally between entangled particles, special relativity is not violated.*

In other words any information that might have been sent faster than the speed light *must have consisted of some degree of unreliability; it is "low fidelity" by its very nature.* Hence it is not the kind of causative influence that is barred by special relativity in space-like cases or that disobeys classical causality. Moreover, the information transmitted subsequently to assemble the data consisted of ordinary messaging, which of course does also not violate special relativity and which is hardly mysterious.

Heeding this subtle but compelling concept, here is an additional postulate for the renowned mystery in entanglement, my COMMUNICATION OF PROBABILITIES POSTULATE:

> *Only probabilities are passed between entangled particles, and the communication of probabilities does not conflict with special relativity.*

The key point is that jiggling—and hence, as noted earlier, the uncertainty principle—make it impossible to obtain precise simultaneous determinations of, say, location and momentum. This applies even if there were some harmony in jiggling of entangled entities; as I already stated, sufficient uncertainty remains, enough to escape the demands of special relativity. The idea of harmony need not be abandoned, but (using the singers analogy) 100% precise harmony between two distant singers cannot be proven.

To reword the argument, a typical experiment on entanglement proves nothing until all measurements from both particles have been compiled and compared. First, how one photon behaves at one filter is a matter of probability, not certainty, and the correlation in a pair of photons is not strictly determined— indeed in Copenhagen terms it does not even exist!—until the pair is studied. Then data from many pairs must be amassed to glean a credible result, and that result will still reveal at least some variability. Thus no firm truth—nothing free of an element of luck—has been learned while ostensibly having violated special relativity or exceeded the speed of light. And please note the parallel: Wave-particle duality resides in the waves of probability; entanglement resides in the entanglement of probability.

This concept is important in interpreting the many studies on the validity of the argument in EPR (please see page 34). The concept implies that EPR would be refuted only if the entire process—ending with the final determinations of the behavior of pairs of particles—transgressed Einstein's criteria of special relativity. Yet as Jones (p. 189-190)[24] avers, "no faster-than-light effects can ever be observed

---

[23] The related math and detailed discussion appears in Bell's theorem in experimental tests. Bachelorarbeit, Univestität Wein (University of Vienna) July 2011, available (2015) online.
[24] See Bibliography starting on page 71.

directly..." That is to say, on the basis of these experiments one cannot claim that non-local causality is real, since no one has observed superluminal transmission of precise causative information from one particle to the other in real time. Though this point might not satisfy everyone, in practice the "action" in "spooky action at a distance" can never totally characterize the experiment. I.e., once the experiment is done, no "action" was proved to be non-local, and nothing was actually observed to be "spooky."

Using numbers again, if the probability is around 90% that the polarization of two entangled photons will turn out to be correlated, that probability is around 90% everywhere. Everywhere, because the basis for probability is randomness, specifically the universal randomness of jiggling. In other words, randomness—which appears as the aforementioned probability—is non-local, but it alone is *not* "hard" unequivocal indubitable information. Given that only a range of likely numbers could have been transmitted, nothing certain was done superluminally. There may be enough particle-to-particle harmony in their in jiggling to account for some correlation, but still only "some." Reusing my analogy, the two separated but entangled singers might sound good together, but we cannot be 100% sure whether they cooperated superluminally, and even if they could and did, what they communicated would not have been 100% definite information. They could never succeed in sending instant "hard" instructions on the note to hit. Worse yet, using the above logic, we cannot learn whether they were in tune until someone listened to (observed, "measured") both of them and then sent a conventional report from which to draw a conclusion. No unequivocal information—such as that the singers were "surely in tune or surely not in tune"—had to travel superluminally.

If the two entangled particles in the above example were separated by many light-years, the probability of that correlation was about 90% within one millisecond, even though "hard" communication in that time span could not have taken place. If the above example was in an actual experiment to test Bell's theorem, the necessarily statistical results (about 90% and not 100%) could not conclusively invalidate Einstein's objections, nor his principles of causality, nor special relativity, nor the EPR arguments. Neither can anyone claim under these circumstances (probability less than 100%) that entanglement must be a non-local process, nor that quantum mechanics in general must be non-local. Moreover we should be aware that here this less-than-100% probability is a matter of principle, not a consequence of the technical imperfections that typically plague complex experiments. Even if ideal methods could be executed, the state of the universe—and the characteristics of jiggling in particular—preclude ideal and perfect results. I like to sum up this way: Nature ensures enough uncertainty to obviate proving non-locality.

Let me conclude this line of thought by acknowledging that the possibly non-local nature of entanglement suggests a method for superluminal physical transmission. This idea looms large in science fiction, but my foregoing argument avers that only probabilities are entangled, making it impossible to send a predetermined message via entanglement with 100% reliability. If the above scenario were part of an effort to "beam" Captain Kirk to another planet, the best possible result is a 90%-probable copy. Please recall (page 28) my use of this analogy for wave-particle duality, but the point in the present context is that even if entanglement were exploited to send objects in the form of bits of data, and even if entanglement is superluminal, the process would necessarily mutilate the information being sent; some uncertainty would always arise.

I will return to this issue in a chapter on quantum teleportation, but this concept is clear simpler scenarios: E.g., I want to use entanglement to send my 10-digit phone number to another continent, and the equipment is good enough to transmit one digit with 99% (0.99) accuracy. Then sending two numbers would be limited to (0.99)(0.99) or 0.9801 accuracy, and a three numbers to about 0.97 accuracy. Transmitting 10 numbers correctly becomes less and less likely (per the math of "joint probability"). I would have to admit that "this is approximately my phone number!" In effect, entanglement cannot be used for such delivery. It entails at least some uncertainty, and thus it does not transgress the rules of special relativity. (Entanglement may be valuable in less demanding technologies where approximations and variability are acceptable, but then special relativity per se is surely not violated.)

The large question here is whether these explanations suffice for resolving the entanglement mystery. I mentioned shortcomings in the harmony proposal, such as lack of experimental support, and this harmony does not preclude a superluminal process. The difficulty with the additional idea that no hard information has traveled superluminally is the assertion that "a probability is not an item of hard information." A counterargument could be raised that even a probability (e.g. "about 90%") *can* constitute a hard unambiguous fact. Some interpreters of special relativity might not agree, but that is still an interpretation, one that implies some circular reasoning. E.g., can one legitimately claim that "the fact that the probability of a correlation is about 90%" is not a hard fact? In other words might a probability qualify as a piece of hard information? Given these areas of doubt, I turn to a third postulate regarding entanglement, one with a very different flavor but one which can circumvent the above skeptical considerations.

Here I bring together several ideas. First, Einstein's relativity requires transmission of information through four dimensions (space-time), which is a notion that was not overt in the past. Second, around 1920, once four-dimensionality had been accepted into scientific thinking, Einstein, Kaluza, Klein and others questioned whether it is essential to stop at four dimensions; why not five or even more? [25] Third, a number of modern theories of physics, including string theories, require more than four dimensions, indeed 11 or more dimensions, even if our senses and our experimental methods are oblivious to the additional ones. (These "strings" can be imagined as vibrating one-dimensional strings of energy.)

String theories have a number of attractive and unifying features, but here is the pertinent point: Is it possible that apparently separated entangled particles are local with respect to each other *along one or more of those unseen spatial dimensions*? All that is required is at least one "hidden" or "curled-up" dimension, and this concept—despite its outlandish notions—incurs no mathematical or geometric prohibition. [26] It is hard to visualize but possible to calculate. As I stated, according to the bulk of current experimental experience, quantum mechanics seems to be non-local, so that entangled particles can behave as if there were no space between them. Could it be that *along some dimension(s) there indeed is "no space" between them*? Information could then pass instantaneously along one or more of such exotic dimensions, even though we are not aware of this, making quantum mechanics appear non-local without discrediting special relativity.

Conversely, do very distant particles have apparently space-like separations merely because this space exists along familiar dimensions, ones that are not "curled up?" The stretch of imagination is not prohibitive, and it can be easily worded: Not all dimensions "uncurled themselves" after the big bang; length, height and width obviously did so, but at least one other spatial dimension did not, and it remains "curled up" so tightly that it has eluded detection. In the spirit of EPR, we could then say that special relativity and quantum mechanics are *both* incomplete. Quantum mechanics appears to have hidden variables simply because relativity's space-time has hidden dimensions.

Accordingly I envision that when two particles become entangled (as in the photon source in the previous diagram) *they end up adjacent to each other—they are local!—along at least one such curled-up dimension.* Hence my POSTULATE...

The curled-up dimensions postulate.

I posit that these kinds of dimensions exist, and they allow distant entanglement to act in a local manner. There is a term for the case where one quantum system (e.g. one photon) instantly affects another when the two are very close to each other: quantum coupling.

---

[25] See Greene, pp. 186-203, in Bibliography starting on page 71. One of Kaluza's and Klein's goals was to reconcile general relativity and quantum mechanics. For this goal they proposed an additional (fifth) spatial dimension, and in certain ways this goal has been advanced by string theories.

[26] See Susskind, pp. 339-342, in Bibliography starting on page 71.

We can think of quantum coupling as local entanglement or as entanglement on a quantum distance scale. We only have to picture the systems as very close to each other—in effect "local"—along at least one hidden dimension, so close that (as an analogy) an electric current encoded with information can flow from one to the other. Research[27] has even suggested that entanglement may arise between photons that never coexisted with respect to time rather than with respect to space. This intriguing possibility implies that a hidden dimension may be temporal rather than spatial.

Even though string theories and their varieties are a prominent focus of attention in contemporary theoretical physics, and even though they include the possibility of curled-up dimensions, the obstacle again is lack of experimental support. The notion of hidden dimensions also calls for some questionable reasoning: Can we invoke curled-up dimensions to explain detectable entanglement while claiming that for the same reason we cannot detect these dimensions?

Let me turn to another approach for explaining entanglement, one that is even more speculative and indeed requires large leaps of imagination: A group of theories called "superdeterminism" (suggested by the aforementioned John Bell). We can think of these as the ultimate hidden-variable theories, whose appeal is that they can account for the strong correlations seen in entanglement without implicating superluminal communication between the entangled entities.

Please consider the usual experiments on artificially generated entangled photons. In such experiments the filter settings must be selected randomly, and a presumably independent mechanical source provides such photons. Unbiased selection of filter angles can be delegated to humans, but to be sure we usually use a presumably impartial random-number generator. However, what if the universe—perhaps since its inception—is imbued with an obscure but powerful system of forces that manifests itself in the quantum entanglements that we observe?

Specifically, what if every photon-source and every random-number generator were influenced by a hidden preexisting physical system, *such that in an experiment the angles of polarization and the selected filter angles are correlated enough to deceive us?* Then what looks like instant communication is really a form of very subtle determinism. In other words couldn't nature "conspire" to convince us that we can design bias-free experiments, while in reality the outcomes are predestined by that very nature?

Worse yet, this possibility challenges the concept of free will by signifying that all things, such as us humans as well as our inanimate machines, are subject to such determinism. Thus there can never be pure randomness in the design of our experiments, and all outcomes are invalid because they are naturally forced to be unavoidable. An analogy is two widely separated compasses: when they are "randomly" spun by naïve navigators, both needles promptly and miraculously point north, but that is so only because of the "hidden" Earth's magnetic field. Anyway…

My POSTULATE for this state of the universe is embodied in one word, *superdeterminism.*

This rather exotic possibility has been taken seriously enough to inspire experimentation. A recent very elaborate study was aimed at the idea that photons with completely random polarization can be really obtained (by telescopes) from very distant quasars, since these photons arose from their sources many light-years away and billions of years ago. The premise is that any non-randomness that exists in these photons must be extremely ancient and remote, so how can such photons possibly be biased here and now? These photons were then used in selecting the settings in modern experiments on entanglement, and

---

[27] <u>Entanglement Swapping between Photons that have Never Coexisted,</u> E. Megidish, A. Halevy, T. Shacham, T. Dvir, L. Dovrat, and H. S. Eisenberg. Phys. Rev. Lett. 110, 210403. Published 22 May 2013.

the results concurred that instant communication appears to be is real. This outcome is interpreted as evidence that the existence of superdeterminism as an explanation is exceedingly unlikely. [28]

Clearly, the mystery in entanglement—in short, can nature be non-local—is not easy to assail. However, and to summarize, I have four ways in mind to get around this problem; any one of these or even all may be applicable.

First, once particles behave harmoniously because their jiggling has been sufficiently altered, the entangled state that is observed need not violate locality.

Second, even if experiments suggest superluminal communication, only probabilities are entangled. No hard data is ever communicated by relying solely on entanglement, so that the very rational stipulations of special relativity and ordinary causality are satisfied.

Third, current theoretical physics recognizes the possibility of extra and curled-up dimensions that sanction apparently non-local behavior. Proximity along such a dimensions assures locality.

Fourth, theories on superdeterminism, though farfetched and currently unsupported, may conceivably explain our observations on entanglement without requiring instant communication.

Incidentally, please note the historical irony in the first and second ideas: The probabilistic nature of quantum mechanics, which Einstein forcefully disparaged, prevents useful superluminal communication, which his special relativity rejects. We could say that his decried "spooky action at a distance" turns out to be "uncertain action at a distance."

Which of these approaches do I favor? The first is parsimonious, in that the only added concept is harmony in jiggling, but the details are complicated and may not be compelling. The second approach resides in the requisites of special relativity, and I find it easy to accept a strict definition of what constitutes meaningful information, but this position does rely on a particular interpretation. The third approach based in string theory is radical, and even if it meshes with current thinking in theoretical physics, it is impossible to prove at this time. The fourth is almost preposterous, but it begs the question, why not? It is possible to disprove superdeterminism? Of course nothing bars us from accepting several concepts in explaining entanglement, and any of them might be confirmed in the future, but at the least entanglement need not be thought of as a hopelessly impenetrable mystery.

Let me end this chapter with mention of an updated experimental finding: In early experiments that revealed non-local communication between two particles, the speeds with which the particles moved was not high, which simplified the experimental design. It turns out that if the particles are moving very rapidly during the experiment, even close to the speed of light, non-locality is still demonstrated. This detail clearly augments the strength of the conclusion, as counter-intuitive as it seems. (Non-locality at all Speeds. Science Daily, June 2021, University if Vienna.)

---

[28] Cosmic Bell Test Using Random Measurement Settings from High-redshift [very distant] Quasars. Rauch, D...Anton Zeilinger, et al, Physical Review Letters, 121, 080403, 20 August 2018.

Quantum Teleportation and Quantum Computing: Mysteries?

Quantum teleportation, henceforth "QT," is the relocation of a quantum state, usually to a distant location. This is a relatively new, exciting and growing field of quantum science, but *it utilizes entanglement*. This fact raises the question I now consider, whether QT entails another "mystery." The issue is important as QT appears utile in quantum communication and networking, in quantum computation and cryptography, and in data compression. I say more about quantum computers near the end of this chapter.

The short answer is no, QT is not really a mystery, but I will elaborate for several reasons. First, science fiction depicts misleading scenarios (e.g. "beaming up"), and I wish to clarify what actual contemporary QT can and cannot do. Second, the physics behind real-life QT is a practical example of the conclusions I reached in the chapter on entanglement, particularly the concepts that only probabilities are entangled and that no hard data is communicated by relying solely on entanglement. Third, in this book I purposely omit as much math as I feel I can, but—as an option for interested readers—I show how QT lends itself to comparatively easy mathematical explanations and insights. These in turn underscore my thesis that quantum mechanics need not be mysterious. The algebra for QT may be tedious, but it is not enigmatic.

Let me dispense with QT in science fiction, particularly where people are "teleported." The relocation of all particles of a person (or any large object) would involve a huge number of molecules and/or atoms, not to mention subatomic particles, well beyond current methods that are limited to a few particles. Far more importantly, the fact is that *QT is useful to transfer only items of quantum information—quantum states*—and not particles or matter. Moreover there are no quantum states that can reconstruct a human being; such states only describe a few quantum observables, such as the polarization or spin of subatomic particles. In short, when Kirk commands "beam me up, Scotty," real QT does not even come close.

Quantum particles—their states to be precise—are described by qubits. A qubit is a unit of quantum information, a kind of sum (a "superposition") representing some physical feature of a particle. Individual quantum states, such as an orientation of polarization or a spin, are symbolized by $|0\rangle$ and $|1\rangle$ (footnote page 47). Generic qubits are then labeled $|00\rangle$, $|01\rangle$, $|10\rangle$ and $|11\rangle$. Thus a qubit is a two-state quantum system, and only qubits get to be teleported. E.g., a photon may be polarized in a direction, horizontal or vertical, but only the directions of polarization (states) are teleported, not the photon. Such data may be vital in a practical application of QT as parts of a code, but they are still only data. Think of sending a secret code consisting of colors; you do not transmit colored postcards, only the colors. (Where information is stored in qubits—e.g. quantum computers—electrons, photons et al. are also used.)

As for the methodology of QT, an entangled pair of particles—e.g. photons—is required, and its two members must be shared between the two locations before a qubit can be teleported. Then their entanglement can provide a rapid and secure conduit for the qubit (because entanglement is fragile), though in principle this conduit can be infinitely long. For these reasons this "quantum channel" between the two sites is established first. However this channel has a major drawback: *Given its reliance on entanglement, it inserts an element of probability and uncertainty into QT*. For that reason QT also requires *a "classical" communication link* between the two sites. Via this link—such as a cable, radio, or laser beam—ordinary bits of information must be transmitted in order to supplement each teleported qubit and thus eliminate the uncertainty. Indeed, given the need for classical communication in QT, the term "quantum teleportation" is partly a misnomer; successful QT demands the use of non-quantum methods.

Enough on the generalities. As is customary in the literature of QT, "Alice" and "Bob" are scientists who execute the teleportation of a quantum state provided by "Charlie." Two entangled particles, labeled A and B, are generated for Alice by one of several well-known techniques. These are a "Bell Pair." One particle, A, is kept at Alice's site and the other, B, will be sent to Bob's site.

A third particle, C, initially belonging to Charlie, is brought to Alice's site, again by various methods. This particle is typically said to be in the quantum state $|\psi\rangle$, *which is the state (the qubit) to be teleported.*

Alice will eventually have only two particles at her site, A—the member of the Bell Pair that was not sent to Bob—and C, the one to be teleported. Alice then performs a "Bell state measurement," a "BSM," of the state of A and C. *The BSM randomly yields a qubit describing one of four possible outcomes called Bell states; each outcome*—each an item of information—*occurs with a probability of 1 in 4 (0.25).* We note that the end product of a BSM is just one item of information: *which of four Bell states "pertains."*

At this juncture let me focus on a basic point: There is nothing mysterious or enigmatic in the fact that a BSM yields one of four possible outcomes; two particles in two quantum states, $|0\rangle$ and $|1\rangle$ can link in only four ways. Please think of it this way: A special deck of cards has only kings and aces, and for the sake of this analogy these are labeled $|0\rangle$ and $|1\rangle$. If we shuffle and deal out a pair, there are four possible equally likely outcomes, $|00\rangle$, $|01\rangle$, $|10\rangle$ or $|11\rangle$. It's that simple, though I stress that $|0\rangle$ and $|1\rangle$ refer to quantum states (such as spin) and *not* particles. However each state represents one item of information.

As a result of Alice's BSM, and by virtue of the earlier entanglement of two particles—e.g. photons A and B—the qubit at Bob's site is instantly forced into one of four possible Bell states, the one that "pertains." E.g. if Alice finds $|00\rangle$, then $|00\rangle$ is the state that pertains. By chance, one of these is identical to the original quantum state $|\psi\rangle$ of particle C, e.g. its spin, and the other three are related but significantly different. However, *Bob does not yet know which of these has reached him,* but Alice is ready to solve this problem: Using the "classical" or "conventional" (non-quantum) channel, Alice simplifies the code for the two-bit pairs listed and sends the information to Bob as 00, 01, 10 or 11. This is a comparatively lengthy step in QT, given the speed-of-light limit in classical communication. For example Alice may phone Bob to give him one of these pairs of numbers, or she can send him an email.

Here I emphasize additional points: The immediate outcome of Alice's BSM is not a "hard" fact but a 1-in-4 probability; please see my discussion on page 41. Therefore Alice's two-bit message cannot be conveyed by entanglement; it cannot be sent reliably via the quantum channel. However, upon receiving the classical-mode message, Bob can modify the qubit for his particle B in one of three ways, or not at all if no modification is needed, so that this qubit becomes identical to $|\psi\rangle$, the original state of C. Thus this qubit has been teleported, but its necessary and identity was sent *without invoking entanglement.*

In the next paragraph I will begin an optional more-mathematical treatment of QT, but even at this juncture the reader may agree with my conclusion: Even if it is still a mystery, *entanglement alone does not suffice for reliable QT.* Without the classical channel, Bob does not have enough "hard" information to modify the state of his particle B so that it surely matches the original state of C. The best he could do is guess, and his chances of being right would be 1 in 4 (even under ideal technical circumstances). Ergo, even if the quantum channel can transmit some information faster than allowed by special relativity (a capability that is convenient), QT cannot make use of it option. Additional vital information—a "hard" fact provided by Alice—can only arrive at commonplace speeds by commonplace means. Therefore, *QT itself does not represent another mystery of quantum mechanics.* Without a classical link it doesn't work!

For interested readers, here is some timely math behind the probabilistic nature of QT, though there will be somewhat repetition. Otherwise, please feel free skip to page 51. Again, the quantum states of three particles participate: the states of the entangled A-B Bell Pair, and the to-be-teleported state of particle C.

It's easier to start with the Bell Pair, whose main feature is that its two member-particles must be entangled. After satisfactory entanglement has been achieved—exactly how is immaterial—the quantum states of the particles are correlated; knowing one allows predicting that of its partner, even if far away. A Bell Bair, comprised of particles A and B together, is a two-qubit quantum state, as it has two pairs of

0's and 1's. Each such pair exists randomly in four different states. The entire set is described and represented by four equations (listed vertically) that are the wave functions for a Bell state, written in a generic format and with commonly used symbols:

$$|\Phi^+\rangle = \frac{1}{\sqrt{2}}(|0\rangle|0\rangle + |1\rangle|1\rangle),$$
$$|\Phi^-\rangle = \frac{1}{\sqrt{2}}(|0\rangle|0\rangle - |1\rangle|1\rangle),$$
$$|\Psi^+\rangle = \frac{1}{\sqrt{2}}(|0\rangle|1\rangle + |1\rangle|0\rangle),$$
$$|\Psi^-\rangle = \frac{1}{\sqrt{2}}(|0\rangle|1\rangle - |1\rangle|0\rangle).$$

Algebraically, a Bell Pair is a quantum superposition of four states—two particles in two states, $|0\rangle$ and $|1\rangle$, that can link in four ways (page 46)—as appears inside the four parenthetic terms. When Alice obtains the Bell Pair, and later when she shares the pair with Bob, that pair will be in these four states:

$$|\Phi^+\rangle, \ |\Phi^-\rangle, \ |\Psi^+\rangle \ and \ |\Psi^-\rangle.$$

(The above is a wave function (or wavefunction), a time-honored concept devised by Schrödinger, Born and others to show the quantum state of a particle or system of particles. When a quantum measurement of a qubit is achieved, we say that the wave function "collapses" into one of the two possible outcomes that were in the superposition, each with a probability of ½ or 0.5 or 50%. A format for a wave function is a "ket" ( )) in the widely accepted Dirac "bra-ket" ($\langle...\rangle$) notation.)

After the quantum channel is established—particle B has been sent to Bob—Charlie prepares a qubit, particle C, to be teleported, which is in some unknown quantum state, such as a certain polarization or spin. Alice will have to work with the two particles, C and A (so she must wait for Charlie to bring her C). That particle C exemplifies a simple quantum system that can be labeled $|\psi\rangle$. Its explicit wave function for the state of C is usually written in the following format:

$$|\Psi\rangle_C = \alpha|0\rangle_C + \beta|1\rangle_C$$

Particle C exists in a superposition of two quantum states, $|0\rangle$ and $|1\rangle$. Thus it is a two-state system in a one-qubit state. I stress that the right side of this equation (another wave function) is a sum weighed by α and β. The α and β are probability amplitudes (each a complex number having imaginary factors) where $|\alpha|^2 + |\beta|^2 = 1$. Squaring these amplitudes gives the probability that each will appear on a measurement, which is usually 0.50-0.50; the " = 1" represents "normalization" so that the total probability adds up to 1.0. Please note similar math on page 18. The mathematical need for imaginary numbers has a subtle but noteworthy physical significance; without these, states cannot be distinguished.

Alice, Bob and Charlie do not know the exact quantum state of C, which means they do not know the values of α and β, and these are not revealed by a properly performed BSM. This is because in general divulging a quantum state causes the complete collapse of a wave function. Thus if Alice were to determine the exact value of α or β, (or if a criminal "snooped") the effectiveness of QT would evaporate. Here I re-emphasize that the aim of the BSM is not the determination of a particular angle of polarization or the orientation of a spin of particle C. It is the identification of one of four possible Bell states.

Now Alice begins her hardest task, a BSM. I treat this BSM as two consecutive sub-processes that I call "interaction" and "actual measurement." This distinction is artificial—a BSM is a complicated but brief event—and this fractionation may not work for other approaches to QT, but for clarity it helps to isolate the changes that are induced by the BSM. The separation into sub-processes also illuminates how the algebraic steps accompany the physical events in BSM. As I see it, interaction + actual measurement = BSM; interaction occurs first, and its details are particularly pertinent to my current thesis.

In the sub-process of interaction, Alice allows the two particles at her site, C and A, to interact on a quantum level, while A is entangled with the distant particle B, the one that is now at Bob's site. Usually a series of laser-beam pulses is aligned between the particle C and A so that the beams comingle, or so that they overlap, or so that both particles are struck. The essence of the physical changes in this interaction is the breaking of the previous entanglement between A and B and the creation of a new entanglement of Alice's two particles, A and C. Though intricate, this is actually a well-document event in quantum physics, and its algebra proceeds as follows.

Breaking the interaction into parts, Alice started out with the state $|\psi\rangle_C$ of particle C. Thus, one quantum system that Alice deals with is described by its wave function, the above equation $|\psi\rangle_C = \alpha|0\rangle_C + \beta|1\rangle_C$. Separately, she started with the Bell state $|\Phi^+\rangle_{AB}$ of the Bell Pair, the first member of the quartet on page 47 representing all four equally probable states:

$$|\Phi^+\rangle_{AB} = \tfrac{1}{\sqrt{2}}(|0_A0_B\rangle + |1_A1_B\rangle).$$

For this condition, the explicit joint wave function *for describing all three particles* is

$$|\Phi^+\rangle_{AB} \otimes |\psi\rangle_C = \tfrac{1}{\sqrt{2}}(|0\rangle_A|0\rangle_B + |1\rangle_A|1_B\rangle) \otimes (\alpha|0\rangle_C + \beta|1\rangle_C).$$

The symbols $\otimes$ indicate vector/tensor multiplication, although they can be omitted. On the left side we see the starting quantum states under consideration; A and B are in entanglement, and C is alone. [29] On the right side, the state of the (entangled) state A and B is available to be joined with the state of particle C (to be teleported). We see this joining as a multiplication (hence the $\otimes$ between the two long terms); i.e., quantum particles are "seen" interacting mathematically as they interact physically.

When the right side of this equation is fully multiplied out (in quantum math, probability amplitudes do multiply), the equation for the total three-particle teleportation system becomes

$$|\Phi^+\rangle_{AB} \otimes |\psi\rangle_C = \tfrac{1}{\sqrt{2}}(\alpha|0\rangle_C|0\rangle_A|0\rangle_B + \beta|1\rangle_C|0\rangle_A|0\rangle_B + \alpha|0\rangle_C|1\rangle_A|1\rangle_B + \beta|1\rangle_C|1\rangle_A|1\rangle_B).$$

Please note how each of the four right-side terms includes the states of all three particles (in the order C-A-B but this is commutative, and in any case the right-side subscripts can be omitted). Please also note carefully how the two different C-states are distributed among four terms. The rest of these terms are constructed with two different A-B Bell qubits. In other words, before interaction the entangled A-B Bell two-qubit pair existed in a quantum superposition of four states that already differed from each other in a random manner. Meanwhile particle C existed in a one-qubit state, whereupon it was forced—e.g. by lasing—to interact with and to "scatter" onto the much larger Bell Pair. And yet no actual measurement has been done at this point; the above is a mathematical description for just the interaction.

Rather fancifully, I picture the above as Alice trying so hard to get the state of Charlie's particle into the quantum channel that she resorts to a violent laser beam that "scatters" the two lone states of C's qubit all over the plethora of Bell states. Though the two immediately preceding equations spell out the algebra and reflect the physics of the process, in a sense they are merely fleeting fiction, since in "real life" the measurements ensue practically immediately.

---

[29] The state of C is a superposition of two states, as shown in its wave function. Readers familiar with quantum computers will recognize use of a Hadamard logical gate whereby a definite quantum state is transformed into a superposition of two states with equal probabilities. The measurement can immediately follow this step via a measurement gate.

The point is that Alice and Bob get to share a newly created C-B entangled state. Meanwhile particle A becomes irrelevant. Even more importantly—because of the entanglement in the quantum channel—*Bob's particle B instantly attains the state of C*, including the four-way probabilistic property inherited from the Bell Pair, but excluding any "hard" information.

In this context we encounter statements in the literature that the states of Alice's two particles, namely the initially non-entangled particle C and her remaining particle A, have been "projected" or "imposed" onto the four states of the Bell Pair, or (confusingly) onto one of those states. I favor the former view, and if we interpret the words "projected" and "imposed" as a "scattered," we will grasp the gist as I presented it above. I.e., "projection" is the scattering in the interaction. Moreover, I like to see the term "projection" used to stress this crucial concept: The state of particle C ends up projected onto each of the four Bell states, yielding four possible equally probable outcomes; in one of these outcomes, the original state of C is projected faithfully; in the other three this state is not faithfully rendered, as if the projection has disturbed ("scrambled") the order and/or the algebraic signs of the terms.

Indeed another equation takes over seamlessly, packed with information, and I consider it to be the central equation of QT. It shows that the interaction has progressed, as the three-particle wave function above has been expanded into an even longer four-term superposition. Mathematically, this step is accomplished by certain algebraic substitutions. For example the first of those four, with subscripts added, becomes

$$|0_A 0_C\rangle = \frac{1}{\sqrt{2}}(|\Phi^+\rangle_{AC} + |\Phi^-\rangle_{AC}),$$

which is combined with the four right-side terms of the previous long equation. The full algebra is more tedious than interesting, but the results—which are interesting—appear in the next box:

$$|\Phi^+\rangle_{AB} \otimes |\psi\rangle_C =$$

$$+ \tfrac{1}{2}\,|\Phi^+\rangle_{AC} \otimes (\alpha|0\rangle_B + \beta|1\rangle_B)$$

$$+ \tfrac{1}{2}\,|\Phi^-\rangle_{AC} \otimes (\alpha|0\rangle_B - \beta|1\rangle_B)$$

$$+ \tfrac{1}{2}\,|\Psi^+\rangle_{AC} \otimes (\beta|0\rangle_B + \alpha|1\rangle_B)$$

$$+ \tfrac{1}{2}\,|\Psi^-\rangle_{AC} \otimes (\beta|0\rangle_B - \alpha|1\rangle_B).$$

In preparation for what is to ensue, let us examine the generic presentation (just 0's and 1's for the two states) in the above box. The left of the equal sign recites what Alice started with: a Bell Pair AB and Charlie's particle C. On the right side of the equal sign (below it) again we first see four terms reflecting the new entanglement between A and C generated by the interaction; each term contains...

$$|\Phi^+\rangle,\ |\Phi^-\rangle,\ |\Psi^+\rangle\ \text{or}\ |\Psi^-\rangle,$$

where the Φ's and the Ψ's each have a negative counterpart. (The ½'s in the box make the multiplication work out correctly.) Next in each term we see a the pattern "$\alpha|0\rangle + \beta|1\rangle$" that previously appeared in the wave function for particle C,

$$|\Psi\rangle_C = \alpha|0\rangle_C + \beta|1\rangle_C.$$

Clearly, $\alpha|0\rangle_B$, $\beta|0\rangle_B$, $\alpha|1\rangle_B$, $\beta|1\rangle_B$, and their signs are "scattered around" as four versions in the four terms of the entire equation. However, the subscripts remind us that the terms "$\alpha|0\rangle + \beta|1\rangle$" and its three variants no longer describe the possible states of particle C but of particle B. Of course that will make a huge difference; the finished BSM (with the "actual measurement" done) will not only tell Alice which state pertains at her site, but it will do the same for Bob at his site as well.

Because if its importance, I wish to ensure that this scenario is clear: Let me assume for the moment that Alice has just completed the BSM and thus has ascertained that the first state, $|\Phi^+\rangle$, happens (by chance!) to "pertain." As the above box and the list reveal, this means that Bob's particle B is now in the state $\alpha|0\rangle_B + \beta|1\rangle_B$. But if Alice had instead found, say, $|\Psi^-\rangle$, which is the fourth case, then Bob's particle would be in the state $\beta|0\rangle_B - \alpha|1\rangle_B$. Similarly for the other two cases, thus including all four versions that could possibly pertain.

Hence I wish to describe the basic principle of the process of identification. In the above list *the terms in the parentheses,* for example the first one,

$$(\alpha|0\rangle_B + \beta|1\rangle_B),$$

differ from each other in only two ways: The placement of the α and β may be reversed, and/or the sign between the two terms may be a plus (+ as in this case) or minus (–). Now here is an "aha" moment in the study of the math of QT: Can the locations of α and β, together with the presence of a + or – sign between two terms, constitute sufficient information that Alice can use to make a positive identification of a state? And can she do so without revealing its details? The answers are yes, and in fact the implementation of this concept is quite feasible. For example two quantum logic gates (newer electronic devices that have replaced transistors) can be used to pin down which of the four states pertains; one gate narrows the possibilities down to two, and the other gate points to one of these.

However, the details of these techniques (more than one exists) must be rather intricate and indirect, and it is not worthwhile to recite them for the current purpose, which after all is to show how QT entails probability and how the problem of this probability can be circumvented.[30] That circumvention entails conventional communication so that an essential non-probabilistic "hard" fact is made available to Bob so as to achieve successful QT.

Of course Bob could try a direct measurement on his particle B, or even just guess, but that only has a 25% chance of being correct; he could glean some information, but he would not know if it matches C's original state, and the rest of the information would be immediately destroyed. In reality Bob's odds are even less because particles—as well as quantum states and entanglements—are elusive and often fail to appear where they should. The odds are even smaller if the original state of C were a particular angle of polarization; the probability of a correct direct measurement or guess would be less than 1/100. In any case Bob needs unequivocal non-quantum—hence slow-to-arrive—data, thus negating any potential advantage of the high speeds attributed to entanglement.

On this basis alone—that a *non-quantum* step is essential in the process—*QT does not qualify as another mystery of quantum mechanics.* Yet the steps in QT do not end here. To take care of loose ends let me return to the fact that upon completing BSM, Alice will have to inform Bob on which Bell state she

---

[30] My favorite mathematical description of quantum teleportation appears (March 2016) in the web site lightlike.com/teleport/teletalk.pdf, Introduction To Quantum Teleportation by Carlos Mochon at The Perimeter Institute for Theoretical Physics, 2006. (Mochon uses spin rather than polarization as the state.) Even more minutiae appear in the web site Quantum Frontiers, A blog by the Institute for Quantum Information and Matter @ Caltech, posted on Sept. 17, 2012 by shaunmaguire. The latter expounds upon the math in the seminal paper, Teleporting an Unknown Quantum State via Dual Classical and EPR [abbreviated] Channels, Charles H. Bennett et al., Physical Review Letters Vol. 70, No. 13, pp. 1985-1899, March 29, 1993.

found. (Bob's particle B is also in a certain quantum state but at first he cannot tell which.) For Bob to recover the one pertinent state of C out of the four possible choices, he will perform a properly selected corrective procedure on his particle B. That procedure is a "unitary operation" (it does not collapse the quantum superposition; it is reversible; the information is not lost), which calls for Bob to use one of three quantum logical gates, usually labeled Pauli-X, Pauli-Z and Pauli-ZX. Yes, geometric XYZ axes play a role; please see the above footnote for references, do search on "Bloch spheres" used to depict qubits, and look into rotations around these axes.

Bob's aim is to convert (physically and mathematically) the state of B to the initial state of C if it isn't that already. In general terms, where the subscripts are superfluous, Bob must recover this quantum condition in order to effectuate reliable QT:

$$|\psi\rangle = \alpha|0\rangle + \beta|1\rangle.$$

Going back to the first possibility I described earlier, if Alice had signaled 00, it would mean to Bob that the particle B in his possession happens to be already is in this state as listed on page 46, where the order of the items was 00, 01, 10 and 11.

Here is a table so summarize this procedure, but now I show the states that Bob may have received for B, as well as what action (which quantum logical gate) each case calls for.

Bell state $|\Phi^+\rangle$ has $\alpha|0\rangle + \beta|1\rangle$ so Alice sent bits 00. Ergo Bob does nothing.
Bell state $|\Psi^+\rangle$ has $\beta|0\rangle + \alpha|1\rangle$ so Alice sent bits 01. Ergo Bob uses X gate to flip β and α.
Bell state $|\Phi^-\rangle$ has $\alpha|0\rangle - \beta|1\rangle$ so Alice sent bits 10. Ergo Bob uses Z gate to change $-$ to $+$.
Bell state $|\Psi^-\rangle$ has $\beta|0\rangle - \alpha|1\rangle$ so Alice sent bits 11. Ergo Bob uses ZX gate to do both.

We note that in each case, particle B ends up in the desired original state of C, which was $\alpha|0\rangle + \beta|1\rangle$.

Other approaches appear in the literature with the same goal: successful QT through the recovery of the quantum state of the initial particle at a distant site. The point remains: this recovery is complicated by reliance on entanglement with its unavoidable probabilities, and non-quantum conveyance of "hard" information is requisite for the success of this recovery. However I see no enigmas that are exclusive to or characteristic of QT. I think of the entangled pair of particles (the Bell Pair) as a component of QT which, by the very nature of entanglement, precludes reliable QT. Consequently, the additional steps that are used—notably communication via a "classical" link—to rescue the situation. These are intricate and technically demanding, but they are not mysterious.

To conclude on a historical note, just the title of the seminal paper on QT, *Teleporting an Unknown Quantum State via Dual Classical and Einstein–Podolsky–Rosen Channels,* [31] explicitly reveals three facts: A state is teleported, a classical channel is used, and a quantum channel is used. The last item is even named for the famous EPR paper (pages 34 and 40) that in effect first questioned the limitations of entanglement. Modern versions of QT, dealing with states of various particles (even atoms) still labor under those limitations: QT is a limited means of communication rather than transportation, and it implicates more technology than just quantum entanglement.

At this juncture I add a discussion on quantum computing, as this new and developing technology overlaps with QT, and it may seem mysterious. The heart of ordinary (digital) computing is a simple "binary" kind of logical gate capable of making the "if…then" decision. The process can be made more efficient by allowing "if…then…else," and more "power" is achieved by various combinations of similar

---

[31] Published by C. H. Bennett et al. Phys. Rev. Lett. 70, 1895-1899 (1993). Please see previous footnote.

and/or more complicated binary processes. The crucial hardware is a relatively rugged transistor, but it has limitations: Basically, it deals with a digital-binary choice, usually labeled $|0\rangle$ and $|1\rangle$, corresponding to only two values, 0 (zero) and 1. It cannot be miniaturized beyond some point, its decisions are made one after another, and even though those decisions are made very quickly, it is impossible to speed them up beyond some limit, while many such decisions need to me made in solving complex problems.

Enter the quantum computer, whose very name sound exotic: The "hardware" that holds the information being processed can be one particle, such as an electron, photon or atom. It can deal with a huge number of choices at the same time (i.e., sets of any numerical values between 0 and 1), and it is effective in far more demanding scenarios by presenting very elusive outcomes in much less time. The underlying principle of a quantum computer is that the aforementioned particles obey the laws of quantum mechanics.

Indeed a quantum computer represents a superb example of an application of quantum theory, as such devices actually employ entanglement, quantum probability, and superposition. As in QT, the mathematical description of two or more choices—two or more "bits" on information—contained in one entity is a superposition, each a bit existing with some probability. In this context the quantum state of a particle is a represented by a "qubit." As in a Bell state, two qubits can hold 4 items of information, and n qubits can hold 2^n items. Thus quantum computers working on a few qubits at the same time can quickly deal with huge amounts of information. And as in QT, quantum logical gates (e.g. Pauli and Hadamard gates) are used to elicit that information from qubits by affecting quantum states of particles.

Readers perusing general descriptions of quantum computing may encounter some form of this statement: In an ordinary computer, bits such as $|0\rangle$ and $|1\rangle$ can only have two numerical values, 0 or 1. In contrast, superpositions in quantum computing are said to allow qubits to have "all possible values between zero and one at the same time." This notion is needlessly mystifying: Yes, qubits can hold many values anywhere between zero and one, *but each with a certain probability.* Moreover the total of all the probabilities in a quantum superposition must add up to one, as on page 47 in terms of probability amplitudes. Which probability applies in a specific case becomes evident when the measurement or determination of the particle's quantum state has been made. Thus I reiterate that there is no mystery in a superposition, even as it mathematically describes a qubit in quantum computing; it is still merely a weighted sum, wherein each addend is quantified by a coefficient that represents the one probability for one state. Nothing is strange here: Of course all addends participate "at the same time!"

The well-recognized features of quantum mechanics—as covered in this book—suffice to explain the main advantage of quantum computers, namely the ability to rapidly solve very difficult problems. For instance if we consider entanglements between several particles under examination—whereby their states are instantly coordinated despite distances between them—then the power of quantum computers is clear: Bits of information can be shared by the particles whose quantum states hold the information. While entanglement is harder to explain than superpositions, both emerge in quantum computing while requiring no modifications to my foregoing presentations. I emphasize that these features allow for processing of many qubits simultaneously rather than sequentially (while these qubits still describe quantum states).

However this methodology also presents major technical hurdles, mainly because of the fragility and short duration of entangled states as well as of superpositions. We say that particles which are entangled and whose possible states are in superpositions are "coherent." Unfortunately they can easily suffer "decoherence," which promptly results in the loss of the information. Hence the need for extremely low temperature in the devices (close to absolute zero), meticulous isolation form the environment, and extreme miniaturization, so that computations can be done fast enough, generally in less than milliseconds. Once the final measurement in a quantum-computer session is achieved, the information should be accessible to users in a useful form—the sought-after outcome—and coherence is no longer essential. (Tunneling also plays a role here; more on that later.)

To further dispel the idea that quantum computers rely on enigmatic elements of science, let me expand on a crucial quantum-computing process in some detail, but to avoid additional complexity please imagine a familiar game, chess. This easily exemplifies important problems that beg for solutions in the modern world. Currently the most famous chess computer may be IBM's "Deep Blue," which is still a binary device relying on "brute force:" an exhaustive and systematic search of all possible candidates for a solution by testing whether each satisfies the goal (winning). Again a huge drawback to this methodology is that the device needs significant time intervals for each step, which quickly add up ("explode") when many ever-branching sets of choices arise. This is the case in a typical chess game as well as in the problems to be solved by modern computers.

Please consider the start of a chess game: The first move can be one of 20 legal possibilities, each of which can be countered with 20 legal moves, so that after the first pair of turns, the board can be in any one of 400 setups or "positions." Obviously the number of possible moves even after the first few pairs of turns grows exponentially into many millions and then billions, as each possible position introduces a multitude of new possibilities. Thus even with very fast processing, a long game would be prohibitively slow. (A chess-playing computer can be programed with information on previous likely games so that the device can apply "past experience," but this is another difficult task.)

Human players usually do not undertake an exhaustive binary trial-and-error process at each turn, wherein each possible move with all its possible ramifications is examined and finally each position that is reached is deemed "good" or "bad." Rather, at each turn a skilled player judges which moves are better and which are worse by applying principles of good chess: controlling the center of the board, giving the pieces mobility, protecting them against capture, etc. Of course players can also memorize past similar games and positions, but then any deviation can entail huge numbers of different consequences. The point is that human players do not evaluate most situations by way of sequential series of binary yes-or-no choices, especially when many pieces participate. Instead humans rely on *sliding scales of relatively better or worse*, (ironically that are analog in nature). Hence typical high-level games end with a player resigning or agreeing to a draw long before a check-mate or a clearly decisive position is reached.

The question then is just *how a quantum computer can process information on a sliding scale*, such as the relative merit of many possible chess moves. To this end a convenient way to describe qubits (many ways are available) is by treating quantum states as *vectors* that have magnitudes and directions. A vector describes something, e.g. a current, that has a magnitude (how strong) and a direction (which way). Indeed a very effective "tool" in this context is vector algebra, which students of quantum computing must master, since quantum gates and qubits are efficiently described as vectors. Thus the effects of quantum logical gates in qubits can be calculated by vector multiplication.

Accordingly, considering two qubits in a superposition, each has a magnitude that represents the probability that a particular quantum state will appear on a measurement. That probability also has a "phase." In this setting a phase is a theoretical and abstract quantum-mechanical property roughly equivalent to a vector direction. However, a phase cannot be measured in just one qubit; it is *physically significant and measurable only when the phase of one qubit is different from that of another related qubit*; i.e. when two qubits in superposition are "out of phase" to a lesser or greater extent.

Here I rely on two analogies: One is the dissonance between two notes played simultaneously on a piano. Dissonance does not appear when just a C key is struck, but it is audible if the adjoining C-sharp key is also struck. Furthermore the dissonance varies with the distance between the two keys, corresponding to how out-of-phase two qubits happen to be. Another analogy entails two analog clocks, considering their hour-hands and assuming that they run at different rates: The hands may coincide—the clocks are in phase—and at times they are out-of-phase. The amount of discrepancy increases and decreases (it rises and falls) throughout the 12 hours. Moreover, I assume that the clocks can be adjusted, *thereby altering the extent of "out-of-phase-ness."*

Back to quantum computers and vectors; here we witness a real and well-documented scenario: *When the phases of two related qubits point in the same direction (e.g. both to noon), the probability that a particular quantum state will appear on a measurement is highest, and when they point in opposite directions (e.g. 3 and 9), the probability vanishes.* Readers will recognize a fact I have stressed: Here we have *waves of probability*, and in this context a probability of a state can have a direction called a phase. We have constructive interference by directions that are more in phase and destructive interference by directions more out-of-phase. We also have the probabilities themselves, which is one reason quantum computers routinely provide outcomes that vary somewhat, usually necessitating repeated runs.

The ability of a quantum computer to alter phases resides in certain kinds of quantum logical gates. For example the phases attributed to the spin of subatomic particles can be adjusted by exposing them to electric or magnetic fields in the computer's components. By convention, phases are represented as rotations around the Z-axis (of a Bloch sphere), just as these rotations appear in QT; please review page 51 and note a similarity between how gates are used in the two different settings. Many other approaches have been devised, called "algorithms" bearing the composes name(s), and new methods are being invented. In any case, as holds for designing a strong chess-playing quantum computer, human players must first decide what constitutes relatively better or worse moves. Adapting these data to programing a quantum computer entails interconnecting various complex gates. In the case of chess, each characteristic of a good move must be converted to some kind of gradation or relative value. Then a move that, say, improves control of the center can be given positive points, while a weakening move is given negative points, and the net point-value determines what happens physically inside a selected gate. This process, generally very intricate and labor-intensive, is said to be far more "confusing" than for binary devices.

The culmination in this scenario now falls into place: Since it possible to program a quantum computer using gates to alter phases, *the device can be made to adjust the degree to which qubits can be in or out of phase.* This process in turn *can be made to affect the probability of obtaining a measurement*, one that reflects, for instance, the best move out of the many that are possible at a point of a chess game. All this is being accomplished and surely will be applied to tasks such as long-range weather forecasting or the design of molecules for new pharmaceuticals. In the former case, a slight change in atmospheric conditions at any location can instigate a huge number of different outcomes, *but the most likely one can be identified.* In the latter case, a shift in the location of just one atom in a large molecule can echo through the entire structure of the atoms connected by various interatomic bonds, all obeying the very intricate quantum laws of particle physics. Nonetheless *new molecular configurations can be revealed* and then evaluated for clinical use. In such situations, binary computers are simply overwhelmed by trying to execute an exponentially growing number of binary decisions one by one.

But we have just begun; for instance the recent elucidation of the most likely structure of a three-atom molecule, BeH_2, by a quantum computer is hailed as a triumph of quantum computation.[32] Many projects are in progress toward an ideal quantum computer, one that is capable of solving very complex problems rapidly and securely. That security depends in part on the fragility of entanglement, such that nefarious intrusions can be promptly detected and defeated. However not everyone is comfortable with such progress. For instance, a very intricate password will surely be much easier to crack.

In any case, please note that the above summary has incurred no major enigmas. The underlying ingredients and concepts, notably probabilities, superpositions and entanglement, apply elegantly in quantum computing, engendering no additional mysteries of quantum mechanics.

---

[32] Fatayer et al., <u>Molecular Structure Elucidated with Charge-State Control</u>, Science 12, July 2019, vol 365 pp 142-145.

The Quantization of Energy

So far I have covered what I deem to be the salient mysteries of quantum mechanics, but several additional issues should be covered: quantization, tunneling, the currently accepted interpretation of quantum mechanics and, given their nature and importance, quantum fields. I turn to quantization next.

Quantization of energy does not qualify as a mystery in the same way as the others I have covered. This property of energy—that energy exists in discreet indivisible amounts (quanta)—has been amply explained and verified, and it is widely accepted without much consternation or bewilderment. Therefore it is not necessary to propose a separate postulate for its elucidation. Still, quantization is so important that some comment is in order. (For brevity I refer to quantization of energy simply as quantization.)

Quantization is revealed in two main ways. One, something quantized has a minimum amount. We can say that a cent (a penny) is a quantum of money. Two, something quantized increases or decreases only by minimum increments called "quantum leaps." The least difference or change is limited to one increment. In the latter context, the receipt or a payment of a cent is a quantum leap. Thus, as is observed in quantum physics, when the temperature of an object changes, that rise or fall occurs only in quantum steps, not as a gradual slope. E.g., if a small amount of hot water is added to a cold tubful, in principle the temperature either does not change or it rises by at least one quantum. Such a subtle but mandatory step-wise change in temperature is counter-intuitive to everyday experience as well as to classical physics.

Historically, quantization played a key role in the initial development of quantum mechanics. (Indeed it gives quantum mechanics its name.) The way certain heated objects ("black bodies") radiate energy could not be satisfactorily explained around the end of the nineteenth century; Maxwell's equations predicted the emission of vastly more energy than was found in experiments. Then around 1900 Max Planck proposed how the observations about "black body radiation" could be explained, but that explanation required that *energy be absorbed and radiated only in quanta*.

At first this seemed preposterous, but as evidence mounted in favor of Planck's approach, it became clear that Planck's idea represented a valid, albeit revolutionary, scientific discovery. For instance later Einstein noted the photoelectric effect which entailed the concept of quanta of light now called photons. The magnitude of the spin of photons was found to be quantized at 1 (a simplified unit); hence we say that a photon is a 1-spin particle.

A dramatic early application of quantization was to explain why atoms do not self-destruct. As crystallized by Bohr, the thinking went as follows: In all atoms, electrons exist more or less in orbits around a nucleus. (The modern model for atoms envisions clouds of electrons ["orbitals"] rather than discreet circular paths.) Classical physics and Maxwell's equations predicted that orbiting electrons should quickly use up their kinetic energy and crash into the nucleus, whereupon all atoms should collapse and all matter should vanish. While the details are intricate, the concept of quantization accounts for why such calamities do not occur. In essence, quantization allows a quantum of energy to persist in each atomic electron (as well as in each nuclear proton), enough to keep electrons "in orbit" indefinitely. This idea also interfaces with my concept of persistent jiggling; electrons can never have zero energy, and we can thus envision how constantly jiggling electrons resist complete orbital decay. (I cannot resist the quip: I jiggle, and therefore I am.)

Incidentally, I referred to the 2012 Nobel Prize on page 30, pointing out that it supports my views on wave-particle duality. Another experimental triumph on the part of the winners was a series of elegant demonstrations of the actual quantization of light energy in a small cavity. While the concept is not new, the evidence has never been so directly elicited in the laboratory.

Today we accept that electrons in atoms are clearly and consistently restricted to certain "orbitals," while each orbital is characterized by a certain quantized amount of energy. An electron really does "resist" existing in some circular locations in an atom but not in other locations, and the latter are the orbitals. This means that orbitals are like waves of probability, but in atoms they are orbitals of probability. (Historically, the initial proposal [ascribed to Bohr] is that an orbiting electron has quantized angular momentum which can only be a multiple of 1, 2, 3, etc. This idea evolved into giving electron orbitals their unique quantum numbers.)

In mathematical terms, the energies (and momenta) of an electrons bound in an atom are distributed in certain restricted denominations—quanta—so that when one of Schrödinger's famous wave equations is used to predict the energy in a hydrogen atom, valid solutions only exist whose energies are quantized. These wave equations can then predict the approximate size of a hydrogen atom, representing another early feat of quantum mechanics.

Here I add another point, with which I emphasize how the various concepts I have discussed fit together. In studying atomic structure we see a critical modern concept, one I mentioned above: An electron forms an orbital which is a "probability density cloud" around a nucleus, and if it is presented appropriately it forms a wave. The classical idea of discreet planet-like atomic orbits is incorrect. For instance, an electron in an atom of hydrogen exists as an orbital which, if it were visible, [33] would look like a spherical cloud of approximately measurable size, one that fades away farther from the center. As I pointed out, such a cloud is a kind of wave of probability for finding electrons. It is therefore safe to say that if the locations of an atom's electrons were not subject to probability and uncertainty—if electrons followed fully predicable orbits with fully predictable momenta and locations—then the dire prophecy of pre-quantum physics would materialize, and atoms as we know them could not exist. In effect, probability and uncertainty are essential for atoms to have their observed size and volume. But what assures that probability and uncertainty are constant features of electrons' behavior? Jiggling!

Quantization of energy can also be visualized as a counterpart of quantization of matter. The latter of course is the observation that subatomic particles can be recognized as bits of matter or material (e.g., by their mass and momentum), and certain such particles appear to be fundamentally limited to minimal quantities; electrons, protons, neutrons, photons, et al. In this sense, matter is quantized into indivisible entities. Given that $E = mc^2$—which itself has been amply verified in the laboratory, in nature, and even in warfare—the energy of a quantum of matter is equal to mc^2. That is to say, the equivalent of a quantum of m is a quantum of E, and where m is quantified, so can E. (Please think this way: If a penny is a quantum, so is its monetary value.) For instance in the photoelectric effect, absorption of one photon (a quantum of light) releases one electron. Again, I see no need for a separate postulate for this concept, since it hardly seems mysterious; quantified matter may necessitate quantified energy.

There is another quantum-mechanical process that at first glance appears mysterious but that in my opinion does not require another postulate, nor a lengthy discussion. That process is tunneling...

---

[33] Visualizing actual atoms is to some extent technically possible now.

Tunneling and Feynman's Quantum Mechanics

Tunneling, though important, is also not a major source of consternation in the study of quantum mechanics. If a solid barrier such as a wall is placed in the path of objects, quantum mechanics allows an exceedingly small but real probability that some of these will penetrate the wall and even pass through it without damage to either themselves or the wall. Such an event, also called barrier penetration, can be demonstrated experimentally for certain particles, in which case the probability of penetration is substantial.

Tunneling occurs naturally in alpha radioactive decay—large unstable nuclei emit smaller permeating helium nuclei—and it has practical applications. Without tunneling by electrons certain modern electronic devices would be impossible. Tunneling appears in superconductivity, wherein electrons can flow for a long time without the need for an applied voltage. A device allowing such tunneling behavior is the Josephson junction (which merited a Nobel prize) that is crucial in the construction of quantum computers. On the other hand tunneling places a limit on how small an ordinary computer chip can be, lest undesired subminiature "short circuits" be allowed.

In our context, I think of tunneling as a practical example of probability. The final location of a particle is subject to jiggling, i.e., to random variations possibly far from where we expect, and this means that one or more particles can end up on the other side of a barrier. After all, any barrier is also made of particles which jiggle in a random manner. On the other hand one of the reasons tunneling is rare and fleeting is that it entails some energy. In fact it is an instance of the energy-time uncertainty principle I discussed on page 9, whereby enough energy must be "borrowed" for a moment of time, but each occurrence is subject to the randomness of the jiggling of energy. This idea is crucial in subatomic physics, as I discuss in the chapter on fields.

The term "tunneling" is somewhat misleading, since this quantum effect does not involve any hole or tunnel dug through the barrier. Rather, I think of a defensive row of soldiers; attacking soldiers can sometimes slip through the gaps and appear on the other side of the row, particularly since the soldiers are all "jiggling." Another approximate analogy is molecules of water seeping through a hard but microscopically porous surface. In any case, given the nature of jiggling and what we know about energy and the structure of matter, tunneling during a long-enough time period is practically inevitable.

In covering two-slit experiments, I mentioned that a photon—by virtue of jiggling—might land on the screen by circumventing the barrier and traversing neither slit. Now I describe how a photon might tunnel through the barrier. These possibilities raise the following concept: Can a jiggling particle reach a given location by traversing any one of a multitude of paths? Does jiggling sanction even an infinite number of potential paths? And if so, why does nature apparently strongly favor the paths we find to be by far most likely? For example, why do most photons reach the screen as expected, and why are tunneling and stray paths not readily evident?

These questions are elegantly answered via the concept that many paths are possible but nature strongly favors the most efficient ones. Very detailed and very technical/mathematical explanations abound in the literature, but now I wish to summarize one particular facet of that explanation, one that constitutes Richard Feynman's [34] Nobel-prize-winning approach to quantum mechanics. His system rests on the concept of path integrals.

---

[34] See Bibliography starting on page 71, particularly Greene (p. 108-12), Polkinghorne (p. 66-68), and of course Feynman.

The path integral approach, a.k.a. the sum over histories approach, is now the accepted most complete formulation of quantum mechanics. In this setting it helps to think in terms of "events." A fundamental (indivisible, solitary) event may be a particle moving from point A at time t_0 to point B at some later time t_1. Classically the particle traverses only one and the same path—a predictable trajectory in space and time—and hence this event has only one "history." As envisioned by Feynman, quantum mechanics allows the particle to traverse many potential paths during the same time interval, and therefore potentially this event has multiple simultaneous histories, one per path. Hence the event may have an infinite number of possible paths and histories. Some paths seem absurd—e.g. from a photon-source on Earth to a nearby screen via Jupiter—but all possible paths are included, in theory and in its mathematics. (Please think of a "two-slit" setup with an infinite number of slits; any one of these can be used.)

Please note that it is not necessary to find a particle physically traversing all possible paths and hence actually occupying all locations in the universe at the same time. The theory only requires a possibility of many—and even an infinite number of—paths and locations. Each possibility occupies a spot in a mathematical superposition of "quantum states," and each state has an associated *probability* (a probability amplitude in the equation; please see page 19) for finding a particle in some physical location. This concept coincides with what I have been voicing: For us to appreciate quantum mechanics, particles need not be imputed with mysterious properties, but their whereabouts and motions are governed by probabilities. No one has found the same photon everywhere at once, but there is always a likelihood of finding it anywhere. The location of a particle is not universal, but its location-probability is universal.

Analysts of quantum mechanics may tell us that the pattern of waves and fringes found in a two-slit wave-particle experiment prove that one photon can exist as a superposition of many states, each representing a separate path through such an experiment; the photon went this way and that way at the same time. I invoke a simpler concept, including in the context of path integrals: The waves and fringes indicate interference between waves of the probability of locating the photons; there is a *probability* that a photon took this path or that path, or indeed any path.

This brings me to how the one actual path (the usually observed trajectory; the classical history; the Newtonian path) is selected naturally from all possible paths. Looking ahead, *the general idea is that the likelihood for deviant paths is reduced, whereas the likelihood for realistic paths is raised.* In particular, by way of destructive and constructive interference, certain paths acquire low probabilities, whereas one path—the one with stationary "action" as mentioned on page 6—acquires maximum probability. This is a powerful application of the very important principle of least action; nature favors maximum efficiency, in that objects always tend to move on paths of least action. (Mathematically, the natural observed path of a particle is such that the variation [variance] in its action is minimized. I cover the principle of least action—more precisely, the principle of "stationary" action, on pages 217-233 of my 2014 book "Relativity Math...." Please see Bibliography starting on page 71.)

While events in nature are probabilistic, these probabilities can be calculated. Each history contributes to the probability for their common outcome—the event. In particular, the probability for the event is obtained after adding together the probability of all the histories within that event. Ergo "sum-over-histories." In other words, the likelihood for an event is built from the combined effect of every path; ergo the "sum-over-paths" or "path integral." Thus the most probable paths dominate the event as we ordinarily witness it, as Newton had analyzed, and as the principle of stationary action requires.

Earlier I averred that a particle can traverse many paths anywhere in the universe and that all paths for one event can be simultaneous. My further assertion is that these are "paths of probability," akin to waves of probability. (The paths can also be called "virtual.") In this view, it is the *probabilities* for all paths that can exist and coexist anywhere. Of course the fundamental factor that permits multiple paths is jiggling. A photon in an experiment on Earth could in theory jiggle its way from a source via Jupiter to a nearby screen on the same laboratory bench, as unlikely as that is.

Next I call upon the notion—inherent in the understanding of constructive and destructive interference—that any two paths can be analyzed as waves of probability which can be in phase with each other or which can be out of phase with each other to any extent. The concept of phase differences is crucial in this picture, but the idea of paths being out of phase is the same as that which explains interference in a two-slit experiment (please recall page 23) and is akin to the workings of a quantum computer (page 54).

The phase actually increases at a fixed rate along each path as the phase cycles repeatedly; the final phase also depends on the length of the path, since a longer path—which also must be faster to comply with its time restraint—provides more opportunity for phase gain. Since the phase of the contribution of each path is proportional to its length, one way to visualize the process is that longer and less direct paths readily attain markedly different phases compared with the phases of shorter and undeviating paths.

In effect, straying promotes dissonant phasing. Conversely, in space-time the shortest and most direct paths—the straightest possible ones—are resistant to slight variations between the paths; they have nearly identical lengths, so they tend to be in phase—they tend to be consonant—with each other. This is why, for example, a thin light beam routinely coincides with the shortest distance between two points. It is also why "the shortest distance between two points is a straight line." Incidentally, proving this is a commonly taught exercise in physics and geometry, and the logic underpins the definition of "action," as in "least action" and "stationary action." Hence my earlier statement that nature favors maximum efficiency.

Altogether now: W*hen paths are summated mathematically in the path integral, those out of phase with each other cancel each other by means of destructive interference. In this way the probabilities for most paths*—the "deviant" ones—*are reduced, while in-phase paths are amplified by constructive "interference."* Deviant paths correspond to non-Newtonian highly unlikely histories, and thus non-classical results for large objects are practically impossible. On the other hand, even the remote possibility of non-classical histories is highly significant because it accounts for the quantum subatomic behavior seen and predicted in research on particle physics; *it elegantly explains why we encounter "quantum weirdness"* (Ferris' apt term for irrational behavior in science; see Bibliography starting on page 71 for reference).

I think of Feynman's approach via the following analogy: An infinite number of different flight paths can lead an airplane from New York to London. Of course based on our experience (and according to Newton's formalism) most of these are extremely unlikely if not absurd and practically impossible, e.g. flying from New York to London via Jupiter. Only one flight path, or at least one set of very similar paths, is ideal. Feynman's triumph was to show mathematically that by far the most probable path is the one that agrees best with our experience, and it is the path obeying the principle of least (stationary) action. In other words, our reality—notably our every-day observations rather than the "weirdness" revealed in subatomic experiments—consists of what is overwhelmingly the most likely. And that likelihood emanates from the properties of nature. In short, the best paths survive.

One notable weakness in Feynman's approach deserves mention: In probability theory, as well as in the mathematics of quantum superpositions, an accepted principle is that the sum of all probabilities in a complete system should equal to 1 (or 100%). We say that there should be "conservation" of probability. This concept is not clearly evident in this approach.

I wish to reemphasize that this paradigm applies to both realities. Ordinary reality as it is usually observed agrees with common sense, and yet "weird" quantum reality is mathematically explainable. In any case, given its inherently random nature, jiggling accounts for the multitude of alternate paths or histories that Feynman envisioned, even if most of these seem highly implausible and even if they are extremely improbable—except in the quantum realm!

To close this topic, let us reconsider the two-slit diagrams (on pages 19 through 22) with the following question in mind: Why do we see highest waves at the centers of each pattern? Of course by intuition we would expect this, merely because the distances from the sources to the centers are the shortest. However now we can give a more sophisticated and yet quite rational "quantum answer" with the help of the path integral approach and the principle of least action: The waves of probability fare best and are highest where they are most in-phase, which indeed corresponds to the shortest distances between two points. Even when stated in negative terms, the notion is subtle and ingenious but not mysterious or irrational: what we commonly observe is what is least unlikely, and events most out-of-phase are the most unlikely.

Not so simple are the very influential applications of the path integral approach: quantum field theory and particle physics. (Please recall page 9.) These disciplines deal with the interactions between the elementary (subatomic) particles. Particles participating in such interactions—wherein various particles are created and/or annihilated—essentially follow paths treated by integration, similar to the way I just outlined. The entire topic, particularly is reliance on the concept if fields, is so important that I devote a separate chapter to it, starting on page 62.

Feynman's book <u>QED</u> (see Bibliography starting on page 71) details this approach, particularly in chapters 3 and 4, but an internet search for "Feynman diagrams" will provide many uncomplicated illustrations and shorter discussions of particle physics. I also include selected mathematical technicalities in chapters 4 and 20 of my 2014 book on Relativity Math.., since modern particle physics rests on both quantum mechanics and relativity, which in itself is a fascinating feature of contemporary physics. In that work I review the critical contribution of Dirac, whom I mentioned on page 15, and will I also discuss the mathematical relationship between relativity and quantum mechanics in some detail later. (Also see Bibliography starting on page 71, and in particular please see the cited on-line article by B. Stone which brings ceaseless random "vacuum fluctuations" into the picture; my concepts of jiggling coincide neatly with Stone's non-technical explanation of quantum field theory, and it will resurface in my chapter on fields on page 67.)

Several factors make particle physics a deceptively complicated and difficult subject: Each elementary particle has different physical characteristics, such as mass or energy, charge, spin, range, duration, etc.; the possible interactions—paths, histories—follow complex patterns ("Feynman rules"); and sophisticated mathematics is essential, entailing intricate equations of quantum mechanics and relativity as worked out laboriously by Dirac, Feynman, Schwinger, Tomonoga, and others. Nonetheless, as arduous as it is, *we can calculate the probabilities—in the form of probability amplitudes—of the various interactions between the particles*, and I reemphasize that these are only probabilities. Why are they not certainties? Surely by now you know my answer. It has to do with jiggling.

Technical Micro-Mysteries

Now I turn briefly to two very different "mysteries," ones that appear only in modern research and experimentation on subatomic physics. The setting generally is a high-energy collider or accelerator, such CERN and others around the world. In particular, let us consider the creation of subatomic particles in such a device. We recall a very basic principle if physics, the law of conservation of matter. A logical application of this law in a given experimental setting suggests two findings: When such particles appear, their number (how many) should be fixed, and the types of particles (photons, electrons, etc.) that appear should also be fixed.

Now here are the mysteries we actually encounter: Under these conditions, the number of particles created varies unpredictably, and the types of particles created also vary unpredictably. Such observations astounded scientists a few decades ago, and I call them "technical" because they are not evident except under highly specialized circumstances. Of course they are "micro-mysteries," as they are restricted to the subatomic realm. Indeed these findings are rarely publicized and virtually unknown outside the above context. However, they have turned out to be important in science and have solidified our view about the relationship between special relativity, general relativity, and quantum mechanics. Though I mentioned the concepts already and will return to them in the context of fields, I devote this very short chapter to this topic to emphasize how simple these concepts are in principle and yet how elegantly they cooperate.[35]

To explain these findings, from Einstein's special relativity we apply (from pages 9 and 56)

$$E=mc^2 \text{ and the related equation } E=p^2c^2+m^2c^4.$$

When subatomic particles appear, the process requires energy as is strictly and universally dictated by these equations. However, the amount of energy that participates in such an event fluctuates randomly—in short, it jiggles—as is dramatically evident in the two pertinent forms of the uncertainty principle (page 7),

$$(\Delta p)(\Delta x) \geq h, \text{ and } (\Delta E)(\Delta t) \geq h.$$

Finally the above principles can be adapted to quantum field theory, which anticipates how subatomic particles behave, with two results (treated by Dirac and mentioned on my pages 9 and 60 as well as in my next chapter on page 66):

In terms of the present discussion, *it is virtually inevitable—and certainly no longer mysterious—that the number as well as the type of subatomic particles produced in a collider are not really fixed but will fluctuate randomly.* In terms of the famous controversial long-standing problem in science—the apparent incompatibility between special relativity and quantum mechanics that once baffled the masters of theoretical physics (including Einstein, Bohr, Schrödinger and Heisenberg)—the above conclusion suggests that the issue can be finally settled. *Not only are special relativity and quantum mechanics compatible; they need each other to explain these mysteries.* These observations lead naturally to my next chapter.

---

[35] I cover the topic in some detail in my book on relativity (chapter 20), and McMahon provides a more complete treatment. Please see Bibliography starting on page 71.

The Mystery in Fields

Having touched on many facets of this large and complicated topic, I now focus distinctly on "fields" in the context of in Quantum Field Theory, QFT. The very mention of "fields" and "quantum field theory" drives sophisticated scientists into early retirement. Indeed in the context of physics, just the term "field" often suffers from misunderstanding and bewilderment. Hence here I treat "quantum fields" as another mystery of quantum mechanics (and calling them "fields" suffices for brevity). I prefer to discuss fields by means of different analogies, the "mathematical" and the "physical," as this topic is complicated enough to warrant separate approaches.

My mathematical analogy is a sheet of graph paper. As usual, one vertical line of the graph is the y-axis, one horizontal line is x, and a diagonal line can be called z. We stipulate that everything we do on this sheet must obey certain rules, such as "a straight line is the shortest distance between two points" and "a circle is a curve of constant radius." Very importantly, we heed such rules even if they are based on past observations, derivations, or principles that are tacitly accepted as valid. Thus we accept more complicated rules, such as "$x^2 + y^2 = z^2$," obviously with the intent to diagram the Pythagorean theorem. Please note: *The graph paper acts as a field, and here its purpose is to study, understand and apply a facet of geometry and its attendant mathematics.*

We can then say that this field is a mathematical tool or device that enables us to study, understand, and apply modern science. What is more, in this context we can call the equation "$x^2 + y^2 = z^2$" a "field equation," since it is a mathematical function that characterizes this field. Or said in reverse, field equations adapt this field to our aim, in this case to study geometry. Ergo, if $x^2 + y^2$ did not equal z^2 in this field, the field would be of no use for us. I therefore emphasize a key concept: Mathematical functions, such as this "field equation," ensure that this indeed is a "field" in the context of modern physics. No equations, no field.

The concept is not at all arcane: The intuitive way to study and understand chess or checkers is by means of a field: The 8-by-8 square board is the field, and the rules that apply are how each piece can move. To get a "feel for fields" in science, let us return to our sheet of graph paper: Another rule we can agree upon is that "two parallel lines will not meet." However in doing so, *we must consider the nature of the sheet of paper.* Specifically, we must know something about this field's shape, in this case whether it is flat. Let us further imagine that this field is large enough to be a long athletic field. Now two side-by-side javelin-throwers agree to hurl their javelins down this field in exactly parallel directions. If the earth were flat, then in theory the paths of the javelins would remain parallel no matter how far they are heaved. In terms of Newton's laws of physics, each athlete is accelerating the mass of each javelin in the same direction, meaning that two parallel forces are created, each guiding the javelins' paths. (In the case of flat surfaces, all the above-cited rules hold, including $x^2 + y^2 = z^2$.)

The analysis of the javelin scenario is simple. We apply basic physical equations (e.g. F=ma) and intuitive geometric rules (e.g. parallel lines on a flat surface don't meet). *However, what if the field is so huge that the curvature of the earth affects the javelins' paths?* Worse yet, what if we consider that the earth is not a sphere but approximately an oblate spheroid? And finally what if we admit that sooner or later the javelins will hit the ground because of gravitation? (Newton had surmised these details, but he did not go far enough with them.) Now we see that the better we understand these complexities, the more accurate will be our calculation of what the javelins will actually do. We realize that we are asking two kinds of "field questions:" First, what is the physical nature of the field along the javelins' paths? In this case, what is gravity like here as it relates to nearby objects? Second, what is the mathematical and geometric nature of the field along the javelins' paths? In this case, what is its state of flatness or non-flatness?

Of course in reality we are interested in fields that can be used to study a modern form of science (e.g. subatomic particles and the forces between them), but in that case why consider fields? In reply, let us consider a more physical situation: please imagine a large number of iron atoms near a magnet, and let us recall Newton, who (had he known about atoms around 1680) would have tried to calculate the magnetic forces in three dimensions that affect the behavior of each atom. These forces, like gravitation, would be thought of as "actions at a distance," with no apparent justification for their existence. More to the point, this "classical" approach would be very laborious and inefficient.

Even worse would be calculating the behavior of, say, many snowflakes falling in a winter storm: Still three dimensions, but many forces and factors may be involved, such as gravity, wind, friction, and possibly electro-magnetism (all are actions at distances!) plus—to be vary accurate—the terrain. Classical methods would be overwhelmed, and the math would be prohibitively intricate.

Faraday (around 1830) devised a better way, the field approach, instigating a dramatic shift in thinking about such situations: Each point in an area or space can be described mathematically by a number or value that describes a local physical property, such as the strength of magnetism or gravity at that point, and this description can be expressed in one or a few equations. This approach also avoids questions about what is "action at a distance." Meanwhile the features of the area or space must be considered, such as by equations that define how shape varies point-by-point. Here atoms or snowflakes behave according to the solutions of these equations. I will come back to details about this concept, but in general we can surmise that fields are useful in modern physics because particles as well as forces can be treated more effectively and more easily by field theories.

Thus from the mathematical vantage, a field is a mathematical object with a large or even infinite number of points. These points are mathematical functions that take a particular value—they equal a number—at every point, each quantifying some physical property. I show a very simple example on page 65. An advantage of this approach is that if there are, say, 1,000 iron atoms in the field, *one field equation can describe the direction and strength of magnetism everywhere in that field*, making it unnecessary to solve 1,000 Newtonian force-equations.

Of course one value or number per point may not suffice, so that a vector of two or more values may be needed. For instance one number per point is not nearly enough in three dimensions where there are separate magnetic and electric intensities at each point, each acting in the three directions. (In fact a set of 8 numbers per point would be needed.) Indeed, more than one field may need to be invoked.

Let us fast-forward to Minkowski, who (around 1907) pioneered the concept that still more numbers per point are invoked by considering not three but four dimensions. As I implied, the analysis—even in four dimensions—can be done more easily in terms of fields, and no matter how many atoms are affected, only a few fields need to be considered (two, in the case of electromagnetism). Fortunately, modern math—including calculus, ironically invented in part by Newton—is not daunted by many varying sets of numbers at many points at the same time.

The usefulness of fields can also be readily envisioned for gravitation. Please consider the sun surrounded by nine planets. The Newtonian approach to obtaining a picture of what is happening is to solve the equation for his law of gravitation [36] $F=G(m_1)(m_2)/d^2$ between each of the 10 "bodies," as each moves at a different rate with respect to every one of the 9 others—a colossal algebraic and geometric task. Alternatively, let us imagine a gravitational field with a myriad of locations, each with a certain

---

[36] G is a constant which Newton appreciated, but he could not detect the effects of special and general relativity, given the technology available at that time.

64

amount of gravity, that amount given by a number called a gravitational potential [37]. Now one differential equation, $G_{ik}=-XT_{ik}$ (the key equation for Einstein's theory of general relativity), describes the entire field exactly and completely—also a colossal task but one that is far more feasible and successful by virtue of the field approach. A gravitational field may also be a more satisfying intellectual paradigm than "action at a distance," and even Newton had doubts about the validity of his assumptions.

In 1916 Schwarzschild solved $G_{ik}=-XT_{ik}$ just after Einstein's publication; please see chapter 17 of my book on relativity math. There I stress that the left side of this equation describes the shape of a gravitational field and the right side describes matter (an object) that is in that field; thus the entire equation in effect answers the two "field questions" I posed earlier (in verse order) while envisioning javelins. In fact the gravitational field has the handy feature that its shape graphically reveals how gravitation "works" according to Einstein's conception. This is why descriptions of general relativity typically include a drawing or video of a trampoline-like surface—a field—indented by a heavy ball. Such an object is then seen "falling" or orbiting into that depression whose shape is clearly and intuitively responsible for the path of that object.

Incidentally, even Newton's law of gravitation, $F=G(m_1)(m_2)/d^2$, can be used as a field equation on our Earth, but only as an approximation. Because the effects of relativity on Earth are very subtle (we are a tiny planet), these approximations suffice for most practical purposes, such as in rocketry. That is to say, in the Earth's comparatively weak gravitational field, the set of numbers generated by $F=G(m_1)(m_2)/d^2$ and the set generated by $G_{ik}=-XT_{ik}$ are very similar. However if extreme accuracy is required, as for the GPS (Global Positioning System), the design must take relativity into account, calling upon Einstein's field equations. And when the scenario is something extreme like a black hole, Newton's equations become useless, and the nature of the field—its tightly curved shape—becomes critical.

This concept restates what fields are, at least from the vantage of mathematics: collections of numbers governed by field equations. Then how the sets of numbers vary point to point corresponds to how the objects in the entire field behave. Moreover, two fields of the same kind—e.g. two gravitational fields or two electromagnetic fields—differ by consisting of different numbers. Thus various fields allow us— relatively easily—to accurately describe and predict various physical events; we can "do" modern science with fields. Though such fields are mathematical abstractions, they are particularly vital to modern real-time subatomic physics, where various different forces act between various different particles in four dimensions. Space can be, and indeed is, "spanned" by more than one field. A good example of a group of fields is a weather map considering the current temperature, pressure, wind, moisture, et al.

At this point I mention a potentially misleading convention: Clearly our graph paper deals in only two dimensions, but particles and forces mathematically occupy three-dimensional space and of course four-dimensional space-time (and even more dimensions than that in some theories). Therefore, perhaps we should be talking about "spaces" rather than "fields" in this context. Nonetheless, the "field" remains the key concept in field theories, representing the historic and dramatic shift in paradigm away from a myriad of particle-to-particle forces to one or a few fields obeyed by particles. (We do talk about "vector spaces.)

To illustrate this idea, let me construct a very simple field on the essentially flat page of this book; this field describes a 2-meter pyramid—a three-dimensional object—in terms of a certain physical property, namely its meters of altitude above the ground at 25 points of the field:

---

[37] The gravitational field (in Newtonian terms) is the gradient of the gravitational potential. Thus how the gravitational potential varies point-to-point is a field.

```
O O O O O
O 1 1 1 O
O 1 2 1 O
O 1 1 1 O
O O O O O
```

Here is a field, and yet all you see is a set of numbers! In fact with just proper labeling, 000000111001212100111000000 could also qualify as a field, representing a real solid object. (If the above numbers refer to various depths rather than heights, and if their relationship to each other conforms to certain complicated geometry, they could describe the indented trampoline I mentioned earlier as an analogy for the gravitational field that is envisioned in general relativity.)

Now I shall go back to some related basic ideas: In the case of, say, temperature, each point of the field can be characterized by one number, an intensity or magnitude or amplitude, and it has no direction; hence temperature can be studied in a "scalar field." In the case of wind, where each point has intensity and directions, we imagine a "vector field." Thus scalar and vector fields can coexist, though one field cannot be both. A collection of simple vectors in a space can make a field, but not every field is a simple vector space. As I will relate later, interactions between particles can be understood as interactions between fields, and—to their huge credit—these fields accommodate the celebrated fact that quantized particles may appear apparently from nowhere and vanish into nowhere.

As I hinted earlier, in the case of a gravitational field a derivation of its mathematical description is quite complicated, requiring tensors. (Simple vectors do not suffice, as the four space-time dimensions are curved, and these curvatures are not uniform; hence we use tensors. In some cases we even use "spinor" fields, having to do with geometric rotations.) The general concept of curved space was introduced by Riemann around 1850 and adapted for gravitation by Einstein around 1910. Thus Einstein's gravitational field can be named a four-dimensional "pseudo-Riemannian" tensor field. The result is the modern "geometric" explanation of gravitation, which has been found to be reliable without exception.

Let us now consider that QFT (quantum field theory) is the modern science of (1) subatomic particles—protons, neutrons, electrons and several others—and simultaneously also of the science of (2) fundamental forces of nature, namely electromagnetism, the strong nuclear force, the weak nuclear force, and gravitation. These particles and forces comprise the "Standard Model" of subatomic physics. With this concept in mind, we can make a series of substitutions in the mathematical analogy, and these will make that analogy more meaningful.

In these substitutions, the "rules" that apply in QFT include the accepted "rules" and teachings of *quantum mechanics* and those of *special relativity;* pertinent details follow soon. The "field equations" for QFT are mainly those of Dirac and Feynman (page 15), and solving them yields values (numbers) that quantify some property of field at a given point. As I said, the scientific-predictive power of fields resides in the fact that fields can be characterized by such equations, and hence, for example, we use Einstein's field equations of general relativity as the modern law of gravitation. Here again the advantage of these fields over Newton's mechanics is dual: Ultimately the appropriate math is easier in principle (working with many objects in one or a few fields), and curved dimensions can be accommodated.

This means that—for the purposes of QFT, modern particle physics, and the Standard Model—a space is filled with many "all-pervasive, interpenetrating" (Brian Skinner's words) fields at the same time. This paradigm is hard to envision, particularly since markedly different fields must coexist, but it does work! I also stress that—except in pictures of analogies—no one who has managed to visualize subatomic particles has also seen a field [38]. Yet at the moment someone observes (e.g. in experiments) how subatomic particles appear, disappear, move or interact, the next step is to invoke a mathematical field in order to study, understand and apply what has been noted.

I now turn to a particular physical analogy for a field, one that is more dynamic and that can be visualized and depicted as a tangible model, akin to pictorial analogies used for general relativity. In the literature various things are used in such analogies: Mattresses, oceans, ponds, trampolines, vats of Jell-O, and others. I urge readers to keep in mind that these are only analogies with strictly limited similarities to quantum fields. Still, I like the analogy with a large lake that has an observable surface, and here the "rules" I alluded to "come alive."

What makes such scenarios, even my visible lake, ultimately unrealistic is that a field must obey the rules, some of which are bizarre and/or may not enjoy fully understood bases. Indeed physicists admit that a common source of these rules is simply past observation; we note how particles and forces are seen to behave, whether or not we understand the fundamental causality. Of course new rules and insights may be discovered at any time, given the amount research that is conducted all over the world. On the other hand some rules used in field theory are already well established by quantum mechanics, relativity and/or ordinary physics, so that we have some idea of "why" these arise. Here are the main accepted rules for fields applicable in QFT, with the help of my physical analogy:

Their surfaces, like my lake, must be pliable and fluid enough to allow propagation of perturbations in all directions; think of slapping the surface of water with your hand or dropping a pebble into it; ripples and waves spread out, so that a distant boat may promptly be forced to sway and bob up and down more than it had before. This concept—this "rule"—is native to quantum mechanics, as I will explain shortly.

However the propagation can be very fast, indeed at the speed of light (the "c" in $E=mc^2$). Thus the distant boat can be affected as fast as relativity allows. Of course the concept that $E=mc^2$ lies at the heart of special relativity. In many instances a related equation, $E=p^2c^2+m^2c^4$, arises (as I show on pages 9 and 61) when momentum and kinetic energy need to be considered.

But even without doing anything to the lake, the surface must have a small random but perpetual fluctuation; it is never still and smooth. This is not because of any wind or subsurface currents; it represents a constant and random variation in energy that always exists at the subatomic level of all matter, and this "rule" vitally represents quantum mechanics. Its randomness includes the height and size of the energy fluctuations, so that it is possible—and indeed inevitable—to encounter rare but large variations, though in general the more vigorous ones are also the briefer. In short, a quantum field is never at rest, and if this sounds like my jiggling and the uncertainty principle in action (page 7), my response is, yes! You may ask, whence this energy? My answer is easy to say but hard to prove: ultimately, the big bang.

Disturbances of the lake's surface represent the effects of a spontaneous or artificial addition of energy to a field, but in either case a key feature of this process is that it is quantized. This means that a minimal amount, a quantum of energy, needs to arise or to be added in order to have an effect. By analogy (though intuitively illogical) a lesser amount will cause no ripples or waves or changes at all. Of course this quantum-axiom stems from the earliest observations and insights in the development of quantum mechanics (around the start of the 20th century), and among the pioneers were Planck and Einstein.

---

[38] To be precise, some animals may sense magnetic fields, and perhaps humans may sense some energy fields, but my point is that field-theory fields are mathematical rather than material constructs, even while they elucidate particles and forces.

This concept is salient considering an electromagnetic field. We now know that electromagnetic energy comes in quantized photons, each of which has a certain amount of energy (expressed as frequency), so it is safe to think of an electro-magnetic field as a collection of various numbers of photons with various frequencies. Incidentally, it is correct that photons may have no rest mass, but they are never at rest!

Similarly, the relativistic gravitational field may be thought of collections of quantized gravitons, particularly since current research suggests that gravitons are real. Still, it is possible that gravitons are unnecessary for the theory to work. For instance, the field equations of general relativity might continue to give correct predictions and result, even if gravitons did not exist. In other words fields populated by particles are attractive notions, but the "beauty" of fields lies in their mathematical powers.

Incidentally, here again is why I invoke different analogies for a field. *Mathematically*, we have a collection of numbers, values and/or equations that recount the physical properties of a particle. *Physically*, we have a collection of very dynamic particles in a field that has various properties. We can say that fields are where the abstract equations of math and the concrete particles of matter meet.

As already mentioned, the most celebrated "rule" that applies to these fields is that mass and energy are interchangeable via the now-obvious equation, $E=mc^2$. In particular, a quantum of energy—analogically a disturbance or "excitation" of my lake that incites the propagation of ripples and waves—can become a particle. In short, since energy fluctuates and energy is mc^2, subatomic particles exist. Moreover, and as mentioned on page 9, particles can vanish by reverting into energy. Given that the inter-conversions between energy and matter occur in fields, and given the random fluctuations inherent in fields, *a quantum particle* (e.g. an electron) *can appear and vanish spontaneously at any time anywhere in a field* (e.g. in an electron field). And whereto does a particle vanish? Into energy.

Please note: Special relativity and quantum mechanics meet elegantly in fields! We can say that in QFT, subatomic particles *must* be equivalent to the energy fluctuations of fields. (The fact that a seemingly inert chunk of mass has enormous "rest mass energy" was one of the discoveries of special relativity, as evinced horrifically when that chunk is uranium inside a nuclear bomb.)

Of course these are quite counterintuitive concepts, given that earlier (e.g. Newtonian) physics tacitly assumed that a vacuum (e.g. in outer space or in a physics experiment) is completely empty and inert. In fact one of the criticisms of Einstein's gravitational field is that he concocted it out of nothing in his imagination in order to make his general theory of relativity appear cogent. Yet thanks to QFT, the accepted paradigm today is that—even in a physical sense—a field does occupy space, that it does contain constantly restless energy, that it is not an inert vacuum as once envisioned, and that its activity results in particles. Moreover we have not only devised a theory of quantum fields, but we have identified two main types of fields, each of which plays a very different but mutually imperative role in explaining what we observe, for instance, in real atoms. These types are the fermionic fields and the bosonic fields, to which I now turn.

Fermionic fields are somewhat easier to envision: Their energy fluctuation give rise to particles called fermions, and these particles include the protons, neutrons and electrons that make up atoms. *In effect, fermionic fields account for all matter (and antimatter)*. Again we note the cooperation within the branches of modern physics: The masses in fermions—which can be measured—are examples of the "m" in special relativity's $E=mc^2$, and at the same time they—and their fields—obey the rules of quantum mechanics, such as quantization. We note that each particle has a field, or said more strongly from the vantage of field theory, *each particle arises from a field*. Thus electrons are the quanta of the electron field, and photons are the quanta of the electromagnetic field.

In this context the other main class of fields consists of the bosonic fields, giving rise to particles called bosons (but no analogous antiparticles) that act as force carriers. I.e., bosons "mediate" the interactions between "experiencing" fermions, accounting for the forces holding atomic nuclei together, making electrons or protons repel each other, etc. To clarify "mediation" please think of bosons and related particles—these being constantly emitted and reabsorbed—as messengers that transmit the effects of fields. E.g., they assure the attraction between protons and electrons (fermions) in an atom is a trade or "exchange" of photons (bosons) between them, while electrons in orbitals stay in adequate separation (repulsion) by absorbing photons.

These interactions obey physical laws such as conservation of intrinsic angular momentum (spin). Thus when an electron, a ½-spin particle, absorbs a photon, a 1-spin particle, the spin of the former can change from $-1/2$ to $+1/2$, as the photon provides the magnitude 1-spin (page 55). The electron thus gains quantized energy and can reach a higher orbital location, and if that electron later falls to a lower orbital location it emits a photon, again a 1-spin particle. Interactions also obey laws about charge, so that for instance a proton has a $+1$ charge since it consists of two "up" quarks each with $+2/3$ charge and one "down" quark with $-1/3$ charge. Such intricate events are depicted in Feynman diagrams; page 60 has citations.

In a wider picture, *the interaction among bosonic fields and their particles account for all the aforementioned "basic forces of nature," namely electromagnetism, the strong nuclear force, the weak nuclear force, and gravitation.* Again, each particle of this type also has a field, or in terms of modern QFT, each such particle arises from a field, and an interaction between particles is actually an interaction between fields. (Electromagnetism and the weak force may constitute one entity. The strong nuclear force is deemed to bind quarks together and, as a residual effect, to bind nuclei together.) The newest addition to bosonic fields is the Higgs field, mentioned on page 5. Higgs particles, while behaving as bosons in other ways, appear to explain the mass of particles more fully, in effect by slowing the motion of these particles. I.e., by interacting with the Higgs field, the fields with no rest mass appear to be massive.

In effect then, per current thinking, all fundamental particles and forces arise from fields! But before we get lost in lofty words, and so that readers may appreciate the terminology, here is a very *very* brief summary of the members of fermionic and bosonic fields and their particles, as assembled in the currently accepted Standard Model. A basic concept to keep in mind is that all space is filled with restless energy, and that energy and mass are interchangeable. Here is a glimpse at the Standard Model, starting with fermions and then moving on to bosons:

The fermions are the leptons and the quarks; they are the particles of matter, as they include the familiar electrons, protons and neutrons.

Leptons include electrons, muons and taus.

Each of these particles has corresponding neutrinos and corresponding antiparticles. (Electrons' antiparticles are positrons; in general fermionic particles and antiparticles are created together and annihilate each other, evincing a form of conservation.)

Quarks make up protons and neutrons, also categorized into six paired names: up and down, charm and strange, top and bottom, and "color" sets. (Different "colors" satisfy the Pauli exclusion principle that forbids two otherwise identical particles in one location.)

Now on to bosons, a.k.a. gauge bosons, the force mediators. They are less familiar outside of particle physics, but without them atoms and elements would not exist. A lepton can be affected by more than one force. ("Gauge" refers to heeding the principle of least action; page 58.)

Bosons for the electromagnetic force are photons; a separate field theory, quantum electrodynamics (QED), deals with these, including interactions between electrons and protons; photons mediate the attraction between electrons and protons, thereby holding atoms together.

Bosons for the weak force are the W+, W-, and Z bosonic particles. The weak nuclear force accounts for radioactive decay (e.g. beta decay), fusion of protons with neutrons (into deuterium ["heavy water"] used in making nuclear weapons), and similar processes. This subtle force affects all fermions, i.e., all matter, and its special field theory is called QFD.

Bosons for the strong force are gluons; another field theory, quantum chromodynamics (QCD), deals with these. Thus a proton or neutron consists of gluons that bind together certain quarks (only in "color" combinations that are neutral). The strong nuclear force holds nuclei together, even though protons repel each other. However the range of this force is short, and hence large nuclei are less stable; nuclear fission occurs when and where the strong force is overwhelmed. Please recall mention of alpha decay (page 57).

Bosons for the gravitational force are gravitons, though how these fit into the Standard Model is still not fully agreed upon. Presumably gravitons transmit the effects of fields to all matter, according to the field theory of general relativity. Hence we experience weight.

Higgs bosons help in explaining the mass of particles (as noted above and on page 5).

In the above discussion I mentioned conservation; let me enlarge on this topic since it supports the gratifying concept that quantum mechanics and general relativity participate cooperatively in nature. The wider idea is that of "symmetries," which form a powerful principle of modern physics. In general, when a law of physics does not change—when it shows "invariance"—despite being studied from a different vantage, we say that we are witnessing a "symmetry." In other terms, since physical events (notably those involving motion and gravitation) are typically described in "frames of reference," symmetry exists when laws are invariant despite their application in different frames of reference.

In our current context, a symmetry exists when the behavior of objects—including falling objects as well as subatomic particles—is the same now and at any other time. I.e., the laws of physics do not change despite a shift (a delay) in time. The subtle but crucial point here is that this symmetry implicates the conservation of energy. Another symmetry is evinced when the event is moved (displaced) in space; the same laws hold, but in these cases the conservation of momentum is implicated. Similarly a "rotational" symmetry is tied to the conservation of angular momentum. (These concepts were developed by the mathematician Amalie Noether [Gribbin, p. 257].)

Specifically, since symmetries between fermions and bosons have been identified, the Higgs boson and gravitation must be included. Thus we witness a tightly unified and cohesive system of particles, forces and physical laws (as also displayed by special relativity and QFT, mentioned two pages back). In keeping with my overall theme, the relationship between quantum mechanics and the rest of science is very intricate and still under study, but it is not mysterious; it reflects accepted rational fundamental principles.

Further descriptions of particle physics become far more complicated, as the particles in the Standard Model differ in mass, electric charge, spin, duration, intra-atomic location, abundance, etc. Such data are shown in graph-like and/or tabular pictorial summaries of the Standard Model that readily appear on searches in Google Images, though they obscure the roles of fields and field-mathematics. Nonetheless, hopefully this section suffices as an overview of fields, rendering them less arcane and less foreboding.

In Summary

In summary, are you...

...mystified by the uncertainty principle? *Everything jiggles unpredictably.*

...mystified by wave-particle duality? *Particles are particles, while waves are only waves of probability. These waves can interfere, but the interference is delicate.*

...mystified by non-local entanglement between particles? *Groups of particles can act in harmony. Even then only probabilities, not hard data, are communicated superluminally. There also may be dimensions along which the particles are local, and superdeterminism may dictate the results of our experiments beyond our control.*

...mystified somewhat by quantification of energy? *The universe as we know it would not exist without energy in quanta. Moreover particles are quantified, while $E = mc^2$.*

...mystified somewhat by tunneling? *Particles can jiggle past each other.*

Three paths are open to you at this juncture.

1. You may have concluded that my system for understanding and envisioning quantum mechanics is not for you; perhaps you are not convinced, or you find it to be a superfluous layer of knowledge. You can just "live with" whatever mystifies you about quantum mechanics. Or, if you are a professional physicist, you can "know" and apply quantum mechanics even if you cannot logically explain all of it. For instance, you may be able to build a faster computer with the help of quantum equations but without resolving its mysteries.

2. You may accept my approach and no longer be mystified by quantum mechanics, and you already know enough about this subject. Further reading on this topic is optional.

3. You may wish to know more about how I apply my premises, and you are willing to delve into many details about quantum mechanics to achieve greater mastery of the subject. But you will have to wait for my next book.

Bibliography and Useful References

David Bohm, Quantum Theory (New York: Dover Publications, Inc., 1989).

Timothy Ferris, The Whole Shebang: A State-of-the-Universe(s) Report (Weidenfeld & Nicolson, London, 1997)

Brian Cox and Jeff Forshaw, The Quantum Universe (Philadelphia: Da Capro Press, 2012).

Richard Feynman, QED (Princeton University Press, 1985).

Richard Feynman et al, The Feynman Lectures on Physics, Volume III (Reading, Mass.: Addison-Wesley, 1965).

Brian Greene, The Elegant Universe (New York: Vantage Books, A Division of Random House, Inc. NY, 1999).

John Gribbin, Q is for Quantum (New York: The Free Press, A Division of Simon & Schuster, Inc.,1998).

Nick Herbert, Quantum Reality (Garden City, New York: Anchor Press/Doubleday, 1985).

R.I.G. Hughes, The Structure and Interpretation of Quantum Mechanics (Cambridge Mass., 1989).

Louis S Jagerman, The Mathematics of Relativity for the Rest of Us (Victoria, Canada: Trafford Publishing, 2001).

Louis S Jagerman, Relativity Math Updated and Revised for the Rest of Us (Available on Amazon.com 2014). Here I discuss the relationship between relativity and quantum mechanics is some detail.

Roger S Jones, Physics for the Rest of Us (Chicago: Contemporary Books, 1992).

Manjit Kumar, Quantum (New York: W.W. Norton & Co., 2008).

J.P. McEvoy and Oscar Zarate, Introducing Quantum Theory (Cambridge, UK: Icon Books, 1999).

Johnjoe McFadden, Quantum Evolution (New York: W.W. Norton & Co., 2000).

David McMahon, Quantum Field Theory Demystified (New York: McGraw Hill, 2008).

Linus Pauling and E. Bright Wilson, Jr., Quantum Theory (New York: Dover Publications, Inc., 1963).

John Polinghorne, Quantum Theory (Oxford, UK: Oxford University Press, 2002).

Leonard Susskind, The Black Hole War (New York: Back Bay Books; Little, Brown & Co, 2008).

Fred Alan Wolf, Taking the Quantum Leap (San Francisco: Harper & Row, 1981).

Gary Zukav, The Dancing Wu Li Masters (New York: HarperCollins, 1979).

A brief non-technical discussion of the role of probability in quantum mechanics and in the uncertainty principle appears in the website http://www.physicsoftheuniverse.com/topics_quantum_probability.html. This 2009 article by Luke Mastin underscores many of my key points, also mentioning their main originators. It specifically stresses the concept of a "probability wave."

The following citations deal with the Nobel Prizes awarded to Haroche, Higgs and Englert in 2012-3:

http://www.nobelprize.org/nobel_prizes/physics/laureates/2012/popular-physicsprize2012.pdf [10/10/2012]

http://www.nytimes.com/2012/10/10/science/french-and-us-scientists-win-nobel-physics-prize.html?pagewanted=all

S. Haroche and J. Raimond, Exploring the Quantum: Atoms, Cavities, and Photons (Oxford Graduate Texts, 2006).

Serge Haroche, M. Brune and J.M. Raimond, Experiments with single atoms in a cavity: entanglement, Schrödinger's cats and decoherence. Phil. Trans. R. Soc. Lond. A (1997) **355**, 2367-2380.

http://www.nobelprize.org/nobel_prizes/physics/laureates/2013/press.html [10/8/2103]

L. Piazza et al, Simultaneous observations of the quantization and the interference pattern of a plasmonic near-field. Nature Communications 6, Article number 6407, March 2015.

The following citation deals with a recent Bell-type experiment said to close all loopholes:

B. Hanson et al, Experimental loophole-free violation of a Bell inequality using entangled electron spins separated by 1.3 km. Nature. Published online 21 October 2015 doi:10.1038/nature15759

The following on-line citation is a very simplified explanation of quantum field theory, but it clearly and elegantly incorporates the concept of ceaseless random fluctuations in quantum fields, a. k. a. vacuum fluctuations:

B. Skinner, A Children's Picture-book Introduction to Quantum Field Theory. August 20, 2015. https://gravityandlevity.worldpress.com/2015/08/20/a-childrens-picture-book-introduction-to-quantum-field-theory.

The final citations contain a fairly comprehensive but informal and illustrated presentation of the Standard Model, particle physics, Feynman diagrams, and the pertinent mathematics:

J. Woithe et al, Let's Have A Coffee With The Standard Model Of Particle Physics. March 2017. IOP Publishing Ltd. Physics Education, Volume 52, Number 3. (Available on-line.)

K. Meissner et al, Standard Model Fermions and Infinite-Dimensional R Symmetries. Physical Review Letters (2018) 10.1103/PhysRevLett. 121 091601

Index